Crash!

The demise of Fossil Foods
and the rise of Abundance

Mark Edwards
Greenindependence.org

Key words.

Food	Green solar	Sustainability	Eflation
Freshwater	Fossil resources	Ecosystems	Hunger
Phosphorus	Global warming	Biofuel	Poverty
Agriculture	Climate change	Energy	Drought
Aquifers	Smoke death	Environment	Fertilizer
Algae	Global awareness	Algaculture	Exports
Genetics	Renewable energy	Biotechnology	Pollution

ISBN 1441474838
EAN-13 9781441474834

CreateSpace
Tempe, Arizona

*Best Science Book – 2009 Independent Publisher Awards
Green Algae Strategy: Engineer Sustainable Food and Fuel

Dedication

To my fabulous life partner Ann Ewen and her passion for creating an abundance of gourmet food, attractive gardens, astounding music and loving support for the untraveled green solar energy path and Sarah Edwards who finishes grace before family meals with "Please God, bless this food, and help people who don't have food get some."

To the Harvard University OPM 30, (Owners, Presidents and Managers) class who have generously shared firsthand experiences traveling globally and meeting people who desperately need sustainable and affordable food and energy – now.

Contents

The Green Algae Strategy Series

Mark R. Edwards

The Green Algae Strategy Series focuses on creating Sustainable and Affordable Food and Energy – "SAFE" production. **The Green Algae Strategy Series** are available for free downloading in color PDF for students, teachers and food and energy policy leaders at http://AlgaeAlliance.com. They are also available on Amazon.com and other retailers. These SAFE production books are used in schools and colleges globally for courses in sustainability, engineering, business, political science, social entrepreneurship, food, water, energy, ecology, environment and world future.

BioWar I: Why Battles Over Food and Fuel Lead to World Hunger. 2007. BioWar I, where food is burned for fuel, must be ended by withdrawal – not of soldiers, but of damaging agricultural subsidies.

Green Algae Strategy: Engineer Sustainable Food and Fuel. 2008. Algae offer solutions for sustainable and affordable food and energy because algae are the most productive biomass source on Earth. *Best Science Book* **– 2009, Independent Publisher Awards**.

Green Solar Gardens: Algae's Promise to End Hunger. 2009. Algaculture in small but beautiful solar gardens and algae microfarms distributed globally will enable SAFE production locally.

Crash: The Demise of Fossil Foods and the Rise of Abundance. 2011. Traditional fossil-based agriculture sits precariously on a foundation of unsustainable fossil resources that will become unaffordable and then will run out. Abundant agriculture is sustainable because it uses plentiful inputs that are cheap and will not run out.

Smartcultures: Sustainable Food despite Climate Change and the Mass Extinction of Fossil Resources, 2011. Farmers may recycle farm wastes to their fields using 360 microfarms. Smartcultures give higher yields by providing bioavailable nutrients at just the right time. Farmers save 40 to 50% by reducing input costs. Smartcultures reduce ecological pollution by 90%.

Preface

A parasite that consumes its host does the most unimaginable thing – it destroys its home and assures its own death. Fossil agriculture's parasitic action on natural ecosystems plus its gluttonous consumption of fossil resources foretells the demise of industrial food production by the time our children reach midlife.

Pundits may puff their hope that surpluses today and new technologies tomorrow will sustain food production through 2050 when our world will have three billion more hungry people. The data presented here suggests a re-examination of the metrics for sustainable food and energy. The precise date of a food supply collapse may be arguable, but not the crash. Before villages, cities and nations die from horrific community starvation; wars will be fought over the remaining remnants of fertile soils, freshwater, fossil fuels and fossil nutrients. We must take decisive and immediate action to avoid catastrophic conflict.

Abundant cheap food provides the foundation for modern societies and their economies. Industrial agriculture leveraged on the fossil resource chain is sustainable only as long as the key fossil resources are available and affordable.

In a global warming environment with diminishing resources, farmers will face enormous price increases before they simply run out of critical inputs for food production. A decade before reserves are consumed, farmers will have to abandon their land due escalating input costs and the availability or affordability of vital fossil resources.

Sensible science argues that we should begin immediately to create an alternative and supplemental food supply. We must act now to avoid food supply shortages that cause a food cascade which operates like a bank run, with fear, hording, price spikes and conflict. When one or more food production inputs become unaffordable, a food cascade will follow and may cause more deaths from malnutrition and starvation than have occurred in all the wars in the history of the world.

Several authors have provided credible warnings describing the food system collapse including Paul Ehrlich, *The Population Bomb*, David and Marsha Pimentel, *Food, Energy and Society,* Lester Brown in *Plan B 3.0: Mobilizing to Save Civilization,* Fred Pearce, *When the Rivers Run Dry*, Elizabeth Kolbert in *Field Notes from a Catastrophe,* Sandra Postel, *Pillar of Sand: Can the Irrigation Miracle Last?* and Jared Diamond in *Collapse: How Societies Choose to Fail or Succeed.* However these sources did not propose a supplemental and alternative food source that can be grown with affordable inputs that will not run out, flourish with climate change and recover, recycle and reuse consumed fossil resources.

We currently have 3.7 billion malnourished people on our planet who desperately need food. Gifting food to communities and countries has not solved hunger because growing, processing and transporting food is too expensive by at least an order of magnitude to be sustainable. In addition, gifting food creates dependency. Gifting food inputs – freshwater, seeds, equipment, fuel, fertilizers and agricultural chemicals – seems logical, until the fossil resource reserves run out or become too expensive for farmers to afford. For many farmers and even some countries, the inputs for industrial food production are already too expensive.

The demise of fossil agriculture comes down to simple physics; the supply of economically recoverable, mined fossil resources will end. Many of the mined resources on which plants depend such as phosphorus, sulfur, copper and zinc have limited reserves that are economically recoverable. Farmers cannot substitute any other element for these vital nutrients and there are no practical synthetic alternatives. When farmers cannot afford or find only one input such as phosphorus; our food supply stops.

Farmers currently apply over 200 million tons of mined agricultural chemicals on their fields each year. Modern food production is sustainable only until the first of many critical resources runs out. We currently flush around half these precious nutrients into our municipal and animal waste streams. Roughly half the residual nutrients erode from fields on wind and water and find their way into waterways, polluting well water, estuaries and oceans. The highly diluted lost nutrients are not economically recoverable with fossil fuels.

Crash! proposes a new form of food production, "Abundant Agriculture" that uses nature's first food production system – green solar energy – to provide sustainable and affordable food and energy, (SAFE), production. Green solar captures and stores solar energy in water-based plant bonds through photosynthesis and represents the simplest and most efficient food production system on Earth. The first energy storage system, algae, promises to provide poor, thirsty, hungry and malnourished people food they can grow locally.

Global climate change threatens the viability of food production so a SAFE system must be climate independent and able to withstand heat and storms. People live all over the Earth, so it must produce food and energy in any geography, attitude and latitude. Agricultural pollution threatens not only food production but the health of all living beings on Earth. A SAFE system should reduce rather than add to air, soil and water pollution. Modern food production depends on monocultures of only eight major crops which jeopardizes the entire food supply because a single vector could wipe out huge areas of production. SAFE production must offer substantial crop diversity.

A sustainable food and energy solution that meets these constraints may seem like science fiction. It is not. Abundant agriculture has not been demonstrated at scale, yet multiple companies will be producing thousands of tons of high energy biomass within the next several years for biofuels, food, feed, fertilizers and fine medicines. At Arizona State University, two field sites have been successfully producing algal biomass for several years. Firms in Hawaii, Israel and China have been producing algal biomass for decades. When scaled production and a modest, yet non-trivial set of other challenges are solved, abundant agriculture can provide a truly Green Revolution with an additional healthy food supply that uses the oldest food on Earth – green solar energy stored in algae – as the foundation.

The proposal here is to neither stop nor replace traditional fossil agriculture but to augment our food supply with highly productive, water-based plants that store solar energy as chemical energy that can be harvested for food, biofuel or many other coproducts. The parallel development of abundant agriculture with Industrial food production makes sense since the two food systems do not compete for precious resources – cropland, freshwater or fossil fuel.

Crash! is divided into two parts; the demise of fossil agriculture and the rise of abundant agriculture. The first chapter examines fossil agriculture, world reserves of selected elements and the predictable events that will occur when fossil resources peak. Chapters 2 – 6 analyze the principle reasons fossil agriculture fails sustainability tests. Each chapter finishes with the constraints required to create a SAFE production system. Chapter 7 begins the discussion of abundant agriculture – green solar energy – using simple inputs that will not run out – sunshine, CO_2 and wastewater.

Abundant Agriculture

Fossil agriculture depends on extractive resources,
Diminishes, degrades and destroys precious croplands,
Drains trillions of gallons of non-replaceable fossil water,
Pumps, transports and burns billions of gallons of fossil fuels,
Pollutes ecosystems with chemicals, herbicides and pesticides,
Amplifies global warming, ecological destruction and water pollution.

Fossil agriculture flourishes in the short term,
But is sustainable for only a few decades because it depends,
On 21 fossil resources that will soon be unaffordable or depleted.

Abundant agriculture uses plentiful and cheap resources,
Yields sustainable and affordable food and energy, (SAFE) production,
And consumes no or minimal cropland, freshwater or fossil fuels.

Abundant agriculture is renewable and sustainable,
For many generations of SAFE producers,
It reduces greenhouse gasses and soil, water and air pollution.
SAFE production may occur at any altitude, latitude or longitude,
Consumes only abundant resources that will not run out,
And inputs that are affordable to all people on our planet.

Chapter 1. The Problem – Fossil Agriculture

Therefore everyone who hears these words of Mine and puts them into practice is like a wise man who built his house on the rock. The rain came down, the streams rose, and the winds blew and beat against that house, yet it did not fall because it had its foundation on the rock. But everyone who hears these words of mine and does not put them into practice is like a foolish man who built his house on sand. The rain came down, the streams rose, and the winds blew and beat against that house, and it fell with a great crash.   **– Matthew 7:24-27**

The Green Agricultural Revolution may turn out to be the worst disaster in human history. Increasing food production ignited a population explosion of four billion people and threatens to devour our planet's natural resources necessary to grow crops. The Green Revolution was misnamed – it should have been named the "Black Revolution." The modern methods used to produce our food supply are neither green nor sustainable.

Extraordinary increases in food productivity were built on a false premise – the eroding foundation of black fossil fuels and mined agricultural chemicals. Food production was leveraged on the unsustainable actions of substituting fossil fuels for labor and explosive expansion of irrigation, fertilizer application and agricultural chemicals in pesticides, herbicides and fungicides.

Experts have rationalized that borrowing fossil water, fertile soils, fossil fuels and inorganic agricultural chemicals such as phosphorus from our children occurs out of necessary, to support our modern lifestyles. The logical flaw is that industrial food production consumes, not borrows, non-replaceable resources that will crash and be gone just when our children most need them.

Magic 21 natural resources

Mining trillions of gallons of fossil water and substituting inorganic (mined) fertilizers for organic nutrients did allow short-term productivity gains. However, this fossil food strategy depends on heavy use of fossil inputs and is sustainable only until about a decade before the first of the Magic 21 fossil resources needed for industrial agriculture runs out, probably by 2040, Figure 1.1.[1]

Figure 1.1 The Magic 21 Fossil Inputs and Nutrients

Primary inputs		Micronutrients		
1.	Fertile soil	9.	Carbon	(C)
2.	Freshwater	10.	Oxygen	(O)
3.	Fossil fuels	11.	Hydrogen	(H)
4.	Seeds	12.	Sulfur	(S)
		13.	Magnesium	(Mg)
Macronutrients		14.	Boron	(B)
5.	Nitrogen (N)	15.	Copper	(Cu)
6.	Phosphorus (P)	16.	Chlorine	(Cl)
7.	Potassium (K)	17.	Iron	(Fe)
8.	Calcium (Ca	18.	Molybdenum	(Mo)
		19.	Manganese	(Mn)
		20.	Nickel	(Ni)
		21.	Zinc	(Zn)

Several fossil resources will become unavailable or unaffordable within the next 30 years. Consider that 33% of our planet's prime cropland has been so degraded it had to be abandoned over the last 30 years.[2] Nearly half of the remaining cropland has been so seriously degraded that farmers must now apply significantly more fertilizer and freshwater to grow crops productively.[3]

Millions of farmers have found their wells dry and their fields fallow. Other farmers have found seeds, water, fuel and fertilizer beyond reach because they are unaffordable. Most farmers have had to take loans to pay for expensive crop inputs. When their crop fails, farmers lose the food they were planning to feed their family as well as their farms, because they cannot repay the loans. When a farmer cannot find or afford just one of the Magic 21 fossil resources, industrial farming ends and the food supply stops.

The "Black Revolution" seemed like such a good idea when it was conceived in 1950, leveraged on cheap fossil resources and plentiful freshwater. It did increase food production but productivity enhancements were built on an eroding foundation of fossil resources. Once farmers made the choice to substitute inorganic chemicals rather than replace soil nutrients with crop residues and manure, industrial production became limited to the availability and affordability of fossil reserves stored in the Earth's crust. Industrial farmers harvest their food crop and lose a majority of their valuable soil nutrients in the animal and human waste streams. Farmers then lose most of the remaining nutrients to erosion; wind and water. Industrial food production not only loses a majority of the applied agricultural chemicals but all the fossil fuel consumed to grow, harvest, process and transport the food.

The structure of fossil agriculture, where each year farmers apply tons of newly mined inorganic chemicals and more fossil water, acts like an accelerating landslide that creates three negative outcomes:

1. Once farmers made the choice to leverage food production with erodible assets by substituting fossil fuels and chemicals for organic farming methods (that replaced nutrients with organic resources), they found themselves trapped with no way to turn back and still achieve a fraction of their food productivity.

2. Continual erosion and degradation of fertile soils combined with increasing pest and fertilizer resistance in industrial crops force farmers to try to compensate by boosting the application of fossil agricultural chemicals.

3. Overconsumption of fossil resources, especially freshwater and crop nutrients creates an accelerating death spiral for fossil agriculture.

The bitter irony of fossil agriculture is that every year farmers will be forced to use more fossil resources to sustain production. They must bear increasingly higher expenses until they go broke or the system collapses. The promise of technology improvements in better seeds, new agricultural chemicals, no-till farming, contour ploughing and more efficient irrigation may extend the life of industrial agriculture, but only for a decade. Modern agriculture is doomed – not because it is not productive – but because the underlying resources that enable it to be so productive are not only limited but their high demand makes the resources continually more expensive. The inability of Industrial agriculture to reclaim, recycle and reuse fossil nutrients limits modern food production to the reserves that can be economically mined and that foreign countries with those reserves are willing to sell at a reasonable price.

An additional food source that can tap surplus human, animal and industrial waste streams as well as the oceans to recover and recycle fossil nutrients and put them back into the food supply would enable agriculture based on abundance. Green solar – farming with water-based plants – offers a new, food source that produces nutritious and delicious food while consuming no or minimal fossil resources (cropland, freshwater, fossil fuels and mined agricultural chemicals.) A sustainable and affordable food and energy, (SAFE), production system will bring about a truly Green Agricultural Revolution.

This new food model may be positioned in the context of food security, the food chain on which all life on Earth depends and food production.

Food security

Food enables life. Food provides the most critical energy security for human societies. Human energy comes from food and food comes from plants directly or the animals that eat plants.

Lack of nutritious food ends life with brutal pain and suffering. In spite of the food productivity gains from the Black Agricultural Revolution, 3.7 billion are severely hungry today – about four times the number in 1950.[4] Globally, 27% of children under age five are significantly underweight, and a child dies of hunger every 3.6 seconds.[5]

Fossil Agriculture

Nature repeats a delicate dance daily in every ecosystem on Earth, balancing food supply with hungry consumers who must compete for their portion. Plants capture and store solar energy, transformed to green plant biomass through photosynthesis. We eat the chemical energy in stored in plants as food. With sufficient nutritious food, organisms thrive and multiply. When food becomes scarce, life forms slow their activities and the weakest die. The first to go are typically the young and old.

Most of the energy on Earth comes from the sun. Light enables life as electrons flow from sunlight and plants synthesize high-energy hydrocarbon plant bonds. Photosynthesis is not only the oldest and simplest energy storage system on the planet but the most efficient. Green solar energy moves up the food chain through multiple trophic levels (steps in the food chain) where it is consumed by microorganisms, fish, fowl or animals or harvested by humans.

Plants are able to capture only about 0.1% of the solar energy reaching Earth. This low photosynthetic efficiency means most solar energy is reflected or absorbed by the environment. Food crops capture solar energy to supply their energy needs and use about 25% of their energy for respiration, 35% for building and maintaining structure, and about 35% for reproduction. Plants store a small surplus of energy that they can use in an emergency or, in the case of food crops, may be consumed by hungry predators.

Food production depends on plants' ability to convert solar energy into stored chemical energy useful to humans and animals. The metric for food productivity is simple – the amount of solar energy captured and converted into usable food per acre or hectare of land. Farmers orchestrate a complex set of variables to produce the highest value product on their land. Typically, farmers succeed through augmenting solar energy with human, animal and fossil energy.

Hunter-gatherers expended roughly 80% of their energy finding and collecting food and firewood.[6] They migrated to find sufficient food supplies. Available food limited population growth. The emergence of early agriculture about 11,000 years ago provided a more reliable food supply from small gardens of grains, vegetables, fruits, roots and nuts. Growing food reduced both the time and energy required to gather food.

For thousands of years, farmers cleared and cultivated the land, planted seeds, weeded and harvested. A single farmer using hand labor had to invest about 500 hours to produce an acre of food grain. The same farmer may be able to farm up to 2.5 acres using hand and animal labor. Lacking good seeds, irrigation and fertilizer, this small production provided only a tiny surplus beyond the farm family.

For many centuries, draft animals provided the necessary additional energy for those lucky enough to afford them by augmenting human labor. However, draft animals did not reduce the required investment in human time to produce a crop because food crops consume an entire growing season – typically about 140 days. Draft animals also required the farmer to plant extra fields of forage crops to feed the hungry work animals.

Farmers discovered that fossil fuels could sharply reduce the physical labor required for food production. Fossil fuels have been used to produce food since the 1920s and have eclipsed the need for draft animals and set the stage for industrial food production that began after World War II. Substituting cheap fossil fuels for human and animal labor makes sense because a single gallon of gasoline produces more power than a horse working at maximum capacity for 10 hours or a human working at 0.1 hp for 40 hours a week for three weeks.[7]

Food chain

The food chain supplies vital energy for every living creature. Humans consume, directly or indirectly, about 40% of all the solar energy flowing through the food chain, which makes the food network critical for human societies.[8] The food chain is robust in its breadth and diversity but has two serious weaknesses for sustaining human life:

1. Should any one of the links in the food chain fail, a single trophic level, the entire food system could collapse. Higher tropic level creatures such as animals and humans are most at risk since we

are at the top of the food chain and are dependent on the sustained success of numerous lower links in the chain.

2. The food chain is inefficient and transfers only about 10% of the energy from a lower level to the next higher. Therefore, huge amounts of biomass must be produced to feed animals and even more to feed humans.

When local food supplies ran low for hunter-gatherers, nomadic tribes simply moved to food. Early farmers produced food for local consumption because transportation was crude, inefficient and expensive. Modern farmers grow food far distant from cities. Roughly half the cost consumers must pay for food occurs from the cost of processing, packaging and transporting food to concentrations of consumers – cities. Disruptions to any of the many fossil resources necessary to grow crops or the supply chains that distribute food will put millions of city dwellers in peril.

Food production

Agriculture consumes 12 billion acres of land globally, with about 70% in pasture and grazing land and 30% in crops.[9] The husbandry of feed and food plants enabled humans to diversify their diets and spend less time insuring their food supply. For 99% of farming history, farmers cultivated their fields with organic production as they:

- Grew a rich diversity of plants (called heritage varieties today).
- Rotated crops to replace organic nutrients, to avoid erosion and to minimize pests.
- Applied carbon, crop residues, animal and green manure for fertilization to replenish the soil nutrients.

Added carbon often came from "potash," which was the practice of burning wood, weeds or crop residues in a pot and scratching the ash into the soil. Animal manure comes from draft and other farm animals that recycle organic carbon. Green manure comes from a cover crop that is ploughed under to replace soil nutrients, add rich organic matter and enhance water retention.

For roughly 11,000 years, agriculture was environmentally benign because farmers relied on natural ecological processes. Crop residues were incorporated into the soil or fed to livestock. Manure was returned to fields where it recycled nutrients and soil organics. Typical mixed production farms with crops and animals were closed, stable and sustainable ecological systems that generated few external impacts. The reason traditional agriculture succeeded for multiple millennia was that mixed production conserved vital nutrients.

Agriculture improved human societies but its Achilles' heel was that food production was dependent on considerable labor, good weather, fertile soils, sufficient freshwater delivered at just the right time and the avoidance of pests. When weather, soils or water failed, communities and sometimes entire civilizations, perished. Before the last people died from community starvation, war and illness decimated the population. History will repeat community starvation as regions and countries run out of critical inputs for modern food production.

Community starvation must be the most agonizing way for humans to die. Families are forced to watch helplessly while their weakest suffer. They see their children and elders in prolonged and excruciating misery while the human body consumes its own tissues in a desperate attempt to sustain energy. The victim's skin changes color and loses elasticity while the stomach loses its ability to digest. Eyes sink in their sockets and victims lose their memory and become weak, fatigued and disoriented. Death usually comes from diseases such as dysentery, pneumonia or heart failure that mercilessly attack the weakened body.

Elizabeth Kolbert in *Field Notes from a Catastrophe* chronicled a list of sophisticated cultures that sustained themselves for hundreds of years and then crashed due to overconsumption of critical resources and climate change, primarily multiyear droughts, such as:

- Tiwanaku, Lake Titicaca in the Andes – crash: A.D. 1100, drought
- Classic Mayan civilization – crash: A.D. 800, drought
- Old Kingdom of Egypt – crash: 2,200 B.C., drought
- Akkadian empire – crash: 2,200 B.C., drought[10]

Fossil Agriculture

Jared Diamond in *Collapse: How Societies Choose to Fail or Succeed* describes similar eco-meltdowns that caused the Anasazi of the U.S. Southwest and the Viking colonies of Greenland to crash.[11] He shows how patterns of population growth combined with drought, over-farming and destruction of natural resources leads to deforestation, erosion and starvation.

Substituting fossil energy for labor

The "Black Revolution" that began in 1950 led to a tripling of food grain production from 630 million to 2 billion tons. The breakthrough came from new plant varieties that when given regular heavy doses of freshwater, fertilizers and protection – from deadly pesticides and herbicides – produced high yields. The highest production increases came early, from 1950 to 1973, as the relatively easy changes were made by adding more fertilizer, more irrigation and better seeds for corn and high producing dwarf varieties for wheat and rice. The dwarf varieties provided better structure that were more compact, much like a short gymnast, that prevented the larger head from falling over.

Farming, fishing and mining are among the most physically demanding vocations and also bring the highest risk for injury, disability and death. Considerable hard physical labor and stamina is required to produce food using hand and animal labor. Farming loses its attractiveness when people have to work so hard for relatively little production. Farmers used their ingenuity and began to industrialize food production by substituting fossil energy for human energy.

In 1940, farmers produced 2.3 calories of food energy for every calorie of fossil fuel energy inputs. Industrial farming substitutes fossil fuel energy for human labor and consumes 10 calories of fossil fuel energy to produce each food calorie – a 23 times increase in fossil fuels, Figure 1.2.[12] Modern farming in the U.S. consumes ten times more energy than it provides to society in food energy.[13]

Substituting cheap fossil fuels, especially gasoline and diesel, for human and animal labor makes sense because a single gallon of gasoline produces more power than a horse working at maximum capacity for 10 hours or a human working at 0.1 hp for 40 hours a week for three weeks.[14] Mechanized grain production on a 600-acre farm in 2009 reduces human labor and time investment to only one hour per acre.[15]

Figure 1.2. Substitution of Fossil Energy for Labor

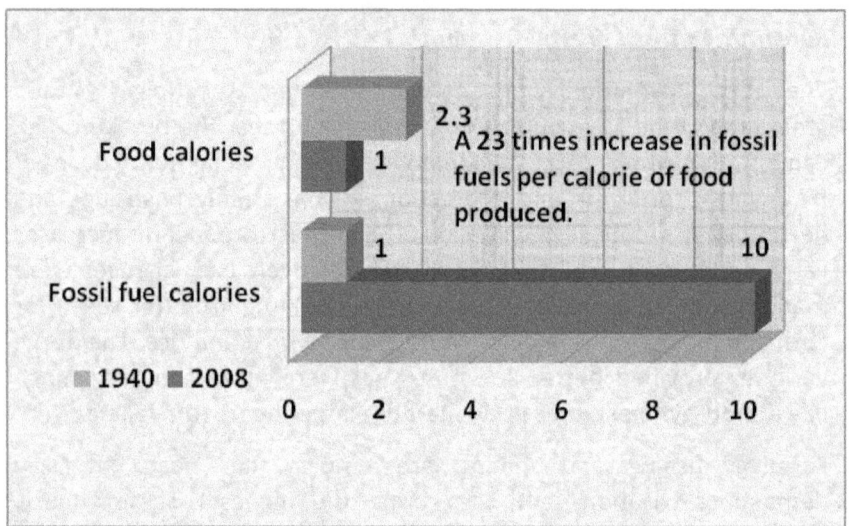

Fossil agriculture

Industrial agriculture is a costly business and fossil resources are the currency used to grow food. After cars, food production consumes more fossil fuel than any other sector of the economy – about 20%. Industrial agriculture depends on fossil fuels for farm machinery, food processing, packaging, transportation, fertilizers, herbicides and pesticides, Figure 1.3. Farmers may cross a field 6 to 9 times on tractors, trucks or harvesters to produce each crop.[16] Tractors pulling plows, disks, cultivators, planters, spray equipment and harvesters consume huge amounts of fuel.

Fossil energy replaces substantial amounts of human labor. For example, assume a gallon of gasoline sells for $3 and an hour of farm labor costs $7. The hour of farm labor has a fossil fuel equivalent of about 200 hours of manpower.[17] This leverage factor has enabled substantial increases in agricultural productivity but there is no safety net should fossil fuels not be available or become too costly.

David Pimentel and Ted Patzek analyzed the fossil energy inputs to U.S. corn production and concluded machinery and fuel, used to reduce human and animal labor, total about 25% of the fossil energy input and the remaining 75% is invested in agricultural chemicals to increase crop productivity.[18] Failing access to fossil resources, modern agriculture could produce only a fraction of current food production because the majority of fossil energy goes to enhance crop productivity.

Figure 1.3 Fossil Inputs Necessary for Modern Agriculture

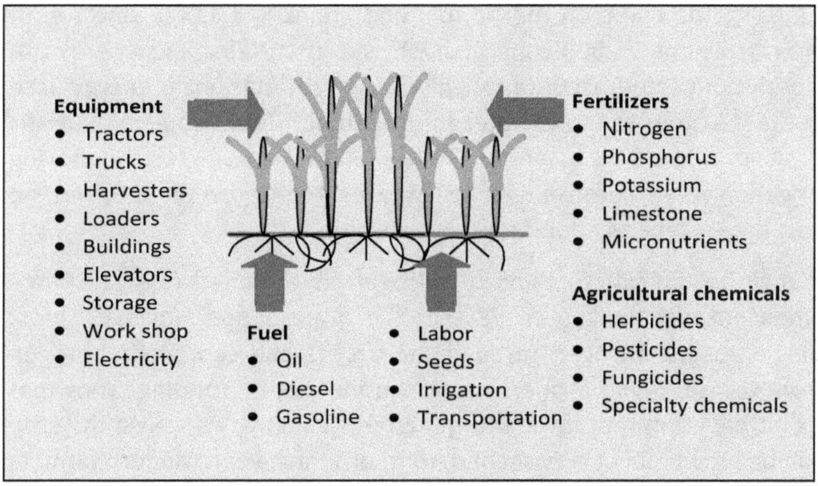

Equipment
- Tractors
- Trucks
- Harvesters
- Loaders
- Buildings
- Elevators
- Storage
- Work shop
- Electricity

Fuel
- Oil
- Diesel
- Gasoline

- Labor
- Seeds
- Irrigation
- Transportation

Fertilizers
- Nitrogen
- Phosphorus
- Potassium
- Limestone
- Micronutrients

Agricultural chemicals
- Herbicides
- Pesticides
- Fungicides
- Specialty chemicals

Farmers put tons of fossil fuel-based herbicides, pesticides and fungicides on crops to control undesired weeds and pests. These agricultural chemicals are produced using extensive amounts of fossil fuels and chemicals. The unintended consequence of expanded use of agricultural chemicals, besides pollution and human health problems, can see be seen in predator and plant resistance figures. In 1950, there were about 10 species of insects resistant to pesticides. Today, there are over 600. Similarly, the number of weeds with herbicide resistance was near zero in 1950 but there are more than 400 today.[19]

Even though insecticide use has increased tenfold, crop loss from insects is double the level it was in the 1940s – about 13%.[20] Pest resistance forces farmers to continually add more chemicals which consume more fossil fuel.

The energy cost of moving water for irrigation is extraordinary since each gallon of water weighs 8.4 pounds. A simple food grain such as corn consumes one million gallons of freshwater per irrigated acre and produces 140 bushels (56 pounds in a corn bushel) or 7,840 pounds. Consequently, it takes about 128 gallons of freshwater to produce a pound of corn. Without fossil energy, water could not be delivered for irrigation and in the U.S., more than half the value of all crops grown are dependent on irrigation.

Globally, the common metric for food grains is 1:1,000; one ton of grain consumes 1,000 tons of fresh water.[21] Water and energy are closely connected because roughly 30% of the stationary energy used in the U.S. goes to move water for irrigation.[22] Few people, politicians or policy makers are aware of the tremendous energy costs in moving irrigation water because water is largely subsidized in the U.S. and the cost data is typically not publically available.[23]

It may be difficult to think of nitrogen as a limited fossil resource when nitrogen makes up 78% of the atmosphere. However, most plants absorb nitrogen through their roots and cannot fix nitrogen from the air. Some farmers use green manure by rotating crops that fix nitrogen but that lowers target crop productivity because the crop can be grown only every second, third or fourth year. Modern farmers favor synthetic nitrogen fertilizer, typically anhydrous ammonia. Nitrogen fertilizer is a fossil input because manufacturing each ton of anhydrous ammonia requires 33,500 cubic feet of natural gas.[24] About 90% of the cost of nitrogen fertilizer comes from the cost of fossil fuels to produce it. American farmers applied over 13 million tons of nitrogen fertilizer to their fields in 2008.[25]

India's success in expanded food production has come at the high cost of a 600% increase in fertilizer use per acre and an even higher expansion of irrigation.[26] Similarly, China has aggressively increased fertilizer and crop irrigation at rates even higher than U.S. farmers.

Fossil Agriculture

About 78% of the corn seeds planted in the U.S. are genetically modified to be more productive but consume substantially more water.[27] Transgenic seeds cannot compete with natural seeds and pests and must be protected with expensive mined chemicals in herbicides, pesticides and fungicides. Seed production, whether genetically modified or hybridized consumes extensive fossil fuels. In addition to the extensive fossil inputs used to grow food, massive amounts of fossil fuels and resources are needed to:

- Drive the tractors and trucks to transport food to processors.
- Convert crops by food processors into forms of food desired by consumers.
- Produce and deliver food additive such as vitamins, minerals, emulsifiers, preservatives and colorings.
- Transport food to retailers who may be thousands of miles away.
- Deliver just-in-time fresh or refrigerated food in trucks.
- Produce food packages and deliver the necessary boxes, cans, paper, plastic, glass, metal and sealing compounds.

Agriculture has supplied humans with food for thousands of years. Yet the substitution of fossil resources for human and animal labor in the last 60 years makes modern food production totally dependent on increasingly scare fossil resources and threatens to annihilate itself. Fossil resource depletion will destroy industrial agriculture. After fossil resource loss, small farms using traditional organic agricultural and considerable labor will be able to produce a little food but probably less than 10% of the amount needed to support existing populations.

Cheap food

The U.S. policy of providing cheap food for citizens and gifting food as foreign aid has turned out to be extremely expensive, unsustainable and damaging to recipients. The politics of cheap food is easy when one third of the food production costs are government subsidized and another third are ignored with no accounting. Many farmers benefit from subsidized fossil fuels, freshwater, electrical power and commodities such as corn. Farmers pay no cost for the resource consumption or ecological damage from food production. Decades of ignoring these costs will leave our children without critical natural resources because food accounting fails to consider:

- Loss of nonrenewable freshwater – 1000 tons of freshwater per ton of grain and about 18,000 tons of water per ton of beef.[28]
- Huge subsidies to big oil that discount fuel to farmers.
- Air pollution and health impacts from growing crops – 2.25 tons of CO_2 per acre, plus nitric oxide.[29]
- Nutrient loss from crop harvest and erosion.
- Water pollution from fertilizers, pesticides and herbicides.
- Dead zones in rivers, lakes, bays, estuaries and oceans.
- Soil erosion – about six tons lost soil per acre per year.[30]

The U.S. foreign-policy operates completely independent of natural resource conservation which threatens to drain America's bread basket dry within two decades. William Ashworth reported in *Ogallala Blue* that Midwest farmers mine five trillion gallons of fossil water a year from the Heartland's aquifer that covers eight states from North Dakota to Texas. The Ogallala aquifer contains water laid down a millennia ago and is not replenished by annual rains. Children of Midwestern farmers will not inherit the valuable farmlands they expect but only dry prairie land that will be practically worthless. They will not be able to grow crops because all the irrigation water was mined by prior generations.

Gifting food provides only a very short term ease of hunger. When the gifted food is gone, people become hungry again quickly. Gifting subsidized food undermines local production. Haiti, for example, has lost nearly all its farmers because cheap subsidized American corn forced Haitian farmers to abandon their land. The farmers could not produce food as cheaply as subsidized American grains. The U.S. subsidies have forced 1.5 million Mexican farmers from their land for the same reason.[31] British International Development Secretary Douglas Alexander said in 2008: "It's unacceptable that rich countries still subsidize farming at $1 billion a day, depriving poor farmers in developing countries $100 billion a year in lost income."[32]

Fossil Agriculture

Fossil resources

Industrial farms are productive because they leverage food production with mechanization, synthetic and inorganic fertilizers, pesticides and herbicides. Fossil farms tend to be so specialized that crop and livestock production are geographically separated. Crop residues and livestock manure that were once recycled have become waste streams that create a disposal problem for farmers. Crop residues are often buried or burned, the cheapest method of disposal. Meat and dairy production occurs indoors or on large feedlots with thousands of birds or animals whose waste pollutes the local ecology.

Macronutrients, called NPK fertilizers, are supplied in large amounts to crops, Figure 1.1 (repeated). Micronutrients also must be available but in lower quantities. Each nutrient serves a specialized role in plant photosynthesis, growth and propagation.

Figure 1.1 The Magic 21 Natural Resources

Primary inputs		Micronutrients		
1.	Fertile soil	9.	Carbon	(C)
2.	Freshwater	10.	Oxygen	(O)
3.	Fossil fuels	11.	Hydrogen	(H)
4.	Seeds	12.	Sulfur	(S)
		13.	Magnesium	(Mg)
Macronutrients		14.	Boron	(B)
5.	Nitrogen (N)	15.	Copper	(Cu)
6.	Phosphorus (P)	16.	Chlorine	(Cl)
7.	Potassium (K)	17.	Iron	(Fe)
8.	Calcium (Ca)	18.	Molybdenum	(Mo)
		19.	Manganese	(Mn)
		20.	Nickel	(Ni)
		21.	Zinc	(Zn)

Fossil fuels receive most the attention in food production because growing food using modern technology consumes huge amounts of energy and many agricultural inputs such as synthetic nitrogen fertilizer can only be made with massive amounts of fossil fuels. Pumping or transporting irrigation water consumes huge amounts of fossil energy. Mining agricultural chemicals, transporting and applying them also consume huge amounts of liquid transportation fuels.

Successful food production requires many other fossil inputs including the specific elements needed for plant growth and development shown in Figure 1.1. Unfortunately, reserves of some of these critical elements will run out long before fossil fuels are completely depleted.

Fossil soil nutrients

Water plants began adapting to early ecosystems 3.5 billion years ago when the distribution of nutrients was very different from today. Plant evolution favored soluble nutrients because they were easy for plants to absorb and transport. All cellular life – plants, animals and humans – use each fossil element for unique cellular functions. For example, phosphorus supports energy transport, ATP, within and among cells, and the plant's DNA and RNA. Phosphorus also supports solar energy absorption and regulation and is vital to the development and maintenance of cell membranes. Failing a single key nutrient, such as phosphorous delivered at the right time, plants die. Gold may be considered a precious metal to man but phosphorus and other fossil elements enable life for plants. Manganese supports photosynthesis and without sufficient manganese, plants are not able to produce sufficient energy to survive. Zinc is critical for all living organisms and supports metabolism and fertilization. Without zinc, plants can neither metabolize their nutrients nor propagate.

Land plants evolved from water-based plants such as algae only 425 million years ago and retained their need for the same nutrients but they needed those nutrients in the soil. Soils remain barren if they lack sufficient nutrients or do retain the soil moisture necessary for nutrient transport through the plant's roots.

Fossil soil nutrients are precious due to scarcity, distribution, reserves and recoverability. Some key elements such as copper, nickel, cobalt and zinc make up only less than one hundredth of one percent of the Earth's crust. Some key elements for fertilizers are found only in a few places on the planet, which means many countries are dependent on the availability and affordability of nutrient imports. As reserves run low, some countries will halt exports or lay heavy tariffs on remaining supplies. As reserves diminish, economic recovery of elements from deeper mines becomes continually more expensive in terms of both dollars and fossil energy.

Growing a food crop for one season removes about 50% of the soil nutrients. Without nutrient replacement organically or with added fertilizer, the next crop lacks critical nutrients and production diminishes. Farmers found by adding more fertilizer and irrigation, they could increase production significantly. However, mined inorganic fertilizers are inefficiently absorbed by plants compared with organic nutrients already in the soil, so farmers must add substantially more fertilizer than the crop actually needs. Higher rates of irrigation act to push more of the applied nutrients into solution where they permeate below the plant's roots or are washed away.

The soil nutrient problem becomes amplified by bioavailability, which means plants need nutrients in a form they can absorb, transport and metabolize. Fossil soil nutrients were laid down over eons and are immediately usable by plants. Applied fertilizers, even manure, are only available to the plant after microorganisms in the soil break down the nutrients. Full breakdown of agricultural chemicals depends on many factors and may take two or more years.

Soil nutrients, especially nitrogen, phosphorus, potassium and calcium dissolve quickly in water which makes them ideal for plant absorption. The downside of the high solubility is that these nutrients are easily rinsed out of topsoil by rain or irrigation water. Soil degradation from erosion forces farmers to apply more fertilizer to obtain the same yield. Erosion and nutrient loss act in combination to reduce yields until eventually the cropland must be abandoned. Global grain yield increases have decreased for the last three decades, largely due to soil degradation.[33]

Fossil nutrient problem

Organic farmers rotate crops with green manure cover crops that replace some of the nutrients lost with harvest. Crop rotation also helps with pest management, builds soil organics and helps with water retention. In contrast, modern farmers replant the same crop, such as corn, each year because they can make more money. They apply inorganic fertilizers to approximate the soil's natural fertility. Unfortunately, both crops and pests develop resistance. Food crops require increasingly more fertilizer to sustain the same level of production. Repeat plantings enable invasive weeds and pests to propagate which creates a significant drag on crop growth. The desire to improve crop productivity combined with rising pest resistance demands more fossil intensive fertilizers, herbicides and pesticides.

Industrial farmers skip the replacement of soil organics because substituting fossil fuels and mined compounds enables significantly higher productivity. Each year a farmer begins preparing the land for a new crop by replacing last year's lost nutrients with additional mined chemicals.

The fossil nutrient problem illustrated in Figure 1.4, occurs because a farmer applies 150 pounds of nitrogen, or other fossil resource, per acre but loses 125 nutrient pounds each year.

Figure 1.4 Loss of Fossil Nutrients – Resource Sink

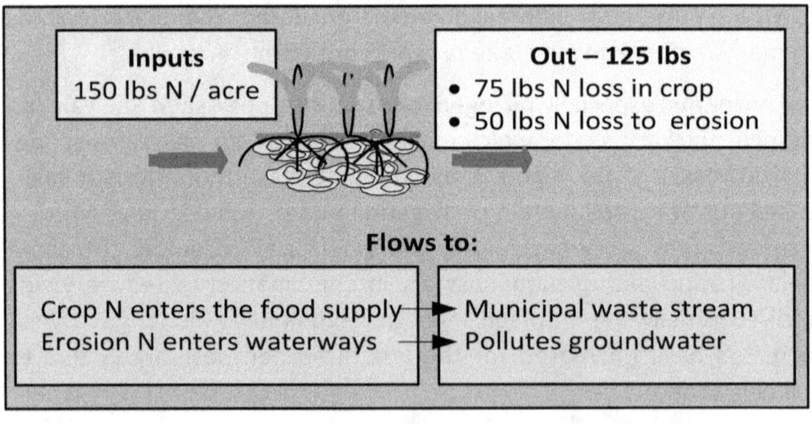

Inputs	**Out – 125 lbs**
150 lbs N / acre	• 75 lbs N loss in crop
	• 50 lbs N loss to erosion

Flows to:

Crop N enters the food supply → Municipal waste stream
Erosion N enters waterways → Pollutes groundwater

The crop harvest removes about half and that nitrogen enters the human food or animal feed chain. Another third may be lost to erosion from wind, rain and irrigation and moves from the soil to nearby waterways where it stays in dilution or is harvested by algae or other water plants. In both cases, harvest and erosion, the nutrients are lost to the field. Each year the farmer begins again with a degraded field and uses fossil fuels to cultivate the field, inflicting more soil erosion and consuming additional fossil inorganic chemicals for fertilizers, herbicides, pesticides and fungicides.

Nitrogen flows into waste streams but occurs in such a low dilution, only a few parts per million, it has practically no bioavailability to food crops even if the water were reused. Land plants simply cannot process enough water to acquire sufficient dissolved nitrogen. The bioavailability problem forces industrial farmers to use mined sources of agricultural chemicals to replace those lost to harvest and erosion. The continual cycle of adding more mined agricultural chemicals makes agriculture an extractive industry that is unsustainable.

Plants cannot be fooled by substituting other elements and there are no synthetic substitutes. Therefore, after mines of phosphorus, manganese, zinc, copper and other vital elements are depleted; the only available source will be recycled waste stream sources or ocean water. Neither option is practical because those sources are too salty for direct use and require far too much energy for economic recovery.

The nature of fossil resources

The nature of fossil resources may be illustrated by their shared characteristics.

Table 1.1 Shared Characteristics of Fossil Resources

Characteristic	Description
Limited supply	When reserves run out, the supply is gone.
Easiest extracted first	The easiest resources, closest to the Earth's surface, are extracted first.
Last half harder to extract	The second half of a reserve may cost 5 - 10 times as much to extract as the first half.

Supply plummets and price skyrockets	Long before the resource runs out, high demand for the limited resource drives prices up exponentially.
Nationalism	Countries with primary sources limit access. Some countries restrict exports – decreasing supply which drives up world prices further.
Regional stock outs	Some regions, especially poor countries, are forced to go without the resource.
No substitutes	Many fossil resources have no effective natural or synthetic substitutes.
No manufacture	Fossil resources cannot be manufactured at all or, in some cases, not at a practical price.
Resource fights	Competitors fight over the last tidbits of each diminishing fossil resource – especially if it means food versus hunger for their family and community.
Fear	Consumers develop a fear mentality that the precious resource will be unavailable to them.
Hoarding	Fear causes suppliers, buyers and greedy speculators, to hoard the resource causing further shortages, more fear and escalating prices.

Access to fossil resources depends on supply, location, cost of production and transportation, environmental consequences, government policies, industry decisions, market price, socio-cultural trends and technological factors. Fossil reserves and years to exhaustion are countable and predictable because mining geologists have explored and mapped the planet for fossil resource reserves that are economically feasible to recover. Two serious factors that are not predictable include a supply disruption and a resource run. Both can have catastrophic consequences for food production.

A supply disruption may occur due to fierce storms that impede mining or transportation, natural disasters such as earthquakes or volcanoes, political actions such as nationalization of mines or trade embargos, terrorist actions or war. Over 30% of the world's phosphorus comes from mines in Morocco and if anything were to happen to those mines, world food supplies would plummet. Since phosphorus is also used for munitions, countries would bid against each other driving up prices of remaining supplies.

Fear of stock outs will lead to a resource run, Figure 1.5, that operates like run a bank where suppliers, farmers and speculators buy remaining stocks and hoard whatever supplies are available in anticipation of even higher prices. A resource run amplifies an already difficult supply situation and has two serious impacts. First speculators and hoarders push prices beyond levels farmers can afford. Second, it removes product from the market which fuels fear and transforms a tropical depression into a category 5 hurricane.

The disturbing detail about a resource run is that a tiny, apparently innocent action can fracture the tripwire and ignite a devastating resource run. The resource does not have to actually be in short supply; just the perception of a supply shortage that fans the flames. The tripwire may be broken by a real or perceived stock out, a supply disruption, government policy limiting resource exports, war, terrorist actions or inclement weather. Once the run starts, rumors proliferate, fear of loss drives people to do stupid things and the cycle intensifies.

Regional stock outs of critical fossil inputs will precede a total end of supply. Poorer countries and poorer farms will be the first impacted by dwindling fossil resources which means regional starvation will occur from local crop failure before mass starvation. However, a single shock from weather, politics or war could interrupt resource supplies and lead quickly to mass starvation because there is far too little food stored to support large populations in mega cities.

Figure 1.5. A Resource Run – Phosphorous Example

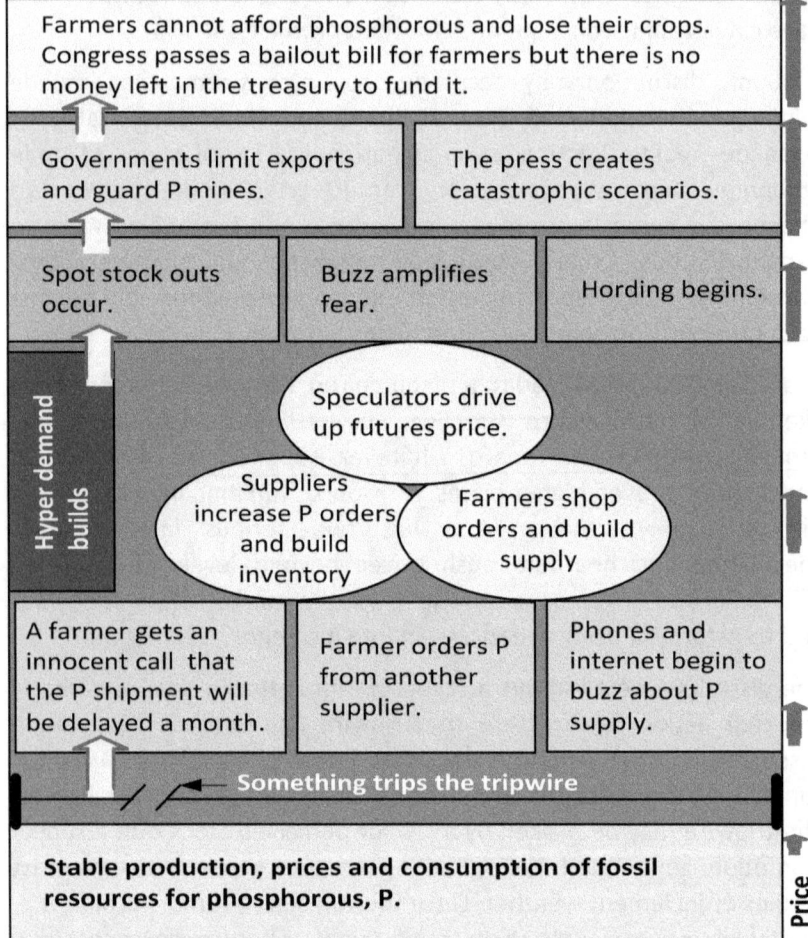

Farmers cannot afford phosphorous and lose their crops. Congress passes a bailout bill for farmers but there is no money left in the treasury to fund it.

Governments limit exports and guard P mines.

The press creates catastrophic scenarios.

Spot stock outs occur.

Buzz amplifies fear.

Hording begins.

Hyper demand builds

Speculators drive up futures price.

Suppliers increase P orders and build inventory

Farmers shop orders and build supply

A farmer gets an innocent call that the P shipment will be delayed a month.

Farmer orders P from another supplier.

Phones and internet begin to buzz about P supply.

Something trips the tripwire

Stable production, prices and consumption of fossil resources for phosphorous, P.

Price

The predicament of farmers in Bolivia provides a case study on spot outages. In 2009, Bolivian farmers were forced to watch their soy beans rot in the fields due to the lack of diesel fuel.[34] Neither farmers nor truckers could find enough diesel fuel to take the crops to food processors. Many farmers were forced out of business, not because they failed to make all the investments necessary to grow their crops, but because government policies created a shortage of diesel fuel.

Bolivia, the poorest country in South America, has banned biofuels because they compete with food and create severe ecological damage. The Bolivian government may join other governments in restricting exports or nationalizing its energy and fertilizer industries.

The demand side of the food supply will create a food cascade where millions of people buy and hoard available food in anticipation of food shortages.[35] If everyone in a nation buys just one extra pound of flour at the same time, stores would have no way to restock their shelves. A food cascade could be ignited by a resource run, food price increases or shortages similar to the riots that jolted over 40 countries in 2008. The food riots were just a prequel to a food cascade. The consequence of a food cascade will be catastrophic and may lead to the loss of 30 million human lives from starvation.[36]

Reserves

World reserves of selected elements illustrate the limitation of fossil resources. At current rates of consumption, world reserves for several fossil resources critical to agriculture will last about 40 years, which may be optimistic. More likely, new technologies that demand more fossil resources will consume reserves at an even faster rate than current production, thereby accelerating the loss of fossil reserves. The Earth Policy Institute headed by Lester Brown used USGS data to calculate global reserves for some important minerals.[37]

Table 1.2. Years Remaining for Recoverable Reserves

Mineral	Global Reserves January 2007	Years Until Reserves Exhausted
	Million Metric Tons	Years
Lead	67	17
Tin	6	19
Copper	480	25

Iron Ore	160,000	54
Bauxite	25,000	68

At least a decade before the reserve for a critical resource is exhausted; farmers will suffer from huge price increases and regional resource shortages or outages. Farmers will endure the same plight as Bolivian farmers because they will have to make the entire investment in their crop and then suffer crop failure from the lack of only one of the 21 magic resources and nutrients. Besides fossil fuels, the most likely candidates for stock-outs include freshwater, fertilizers, pesticides and herbicides, phosphorus, copper and zinc.

Platinum reserves, a key component in pollution controls, catalytic converters and fuel cells, are near exhaustion. Unlike oil or diamonds, there is no synthetic or green alternative. Platinum is a chemical element and when the last unit is used, there will be no more.

Armin Reller, at the University of Augsburg in Germany and his colleagues have been investigating world supplies of metals and estimates that the world will run out of indium by 2015.[38] Indium is used to make liquid crystal displays and has many thin film applications in electronics, biotechnology and manufacturing. Its impending scarcity is reflected in its price which increased 17 times from 2003 to 2006.

Estimating the extractable reserves of many metals is difficult because the figures for rare metals such as indium and gallium are kept a closely guarded secret by mining companies. However, Reller estimates that zinc, which is critical for plant growth and propagation, will be used up by 2037. Hafnium, an important component in computer chips, could be gone by 2017. Terbium, used to make the green phosphors in fluorescent light bulbs, will run out before 2012.[39]

Reserves shown in Table 1.3 do not take into account changes in demand from new products or technologies. While most of these fossil reserves will not impact food production directly, they will create a sense of panic and lead to speculative futures trading and hording of other resources critical to industrial agriculture.

Table 1.3 Fossil reserves of selected Elements[40]

Fossil resource	Years to exhaustion	Fossil resource	Years to exhaustion
Antimony – drugs, semiconductors	15 to 20	**Silver** – jewelry, catalytic converters	15 to 20
Hafnium – computer chips, control rods for nuclear reactors	10	**Tantalum** – cell phones, camera lenses	20 to 30
Indium – LCDs, semiconductors	10	**Uranium** – weapons, energy	30 to 40
Platinum – catalyst, catalytic converters , solar PV cells	15	**Zinc** – galvanizing metals and fertilizer	20 to 30

Water, nutrients and fuel

Modern food production sits precariously on a three legged stool of extracted fossil resources, including especially freshwater, soils and nutrients and fuels, Figure 1.6. When a fossil leg breaks, and it will, food production will crash. The Earth simply has too few fossil resources to sustain a food production system that ravenously consumes precious non-renewable reserves.

Agriculture's own self-destructive actions undermine food production because fossil agriculture is consumptive as it depletes and displaces water, soils, fuels and nutrients. As non-renewable resources near depletion, not only farmers will be unable to support expanding populations, they will unable to support the existing population.

Figure 1.6 Fossil Agriculture's Three Legged Stool

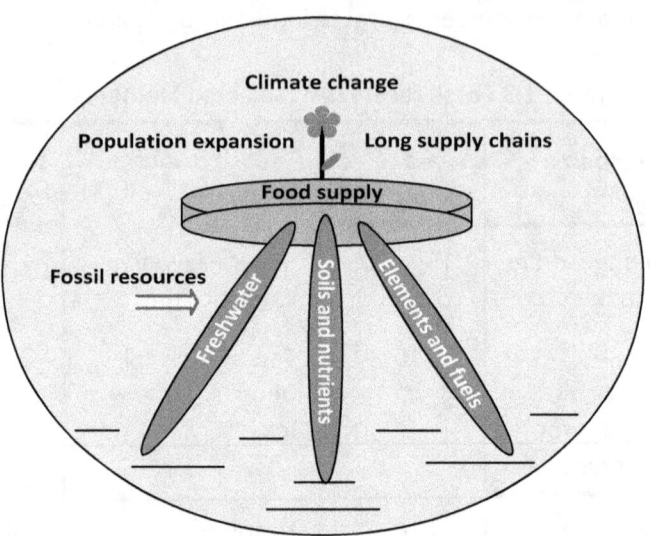

Considerable debate focuses on exactly when reserves of each fossil resource will end. However, there is no debate on whether fossil resources will run out because nearly all have limited supply that can be recovered economically. Industrial agriculture operates in great peril because the food supply fails when only one of the Magic 21 resources becomes unavailable or unaffordable.

Fossil agriculture also faces daunting external challenges including population expansion, climate change, long supplied chains, social and political impacts and monocropping. The next section examines the most critical fossil resource, freshwater.

Chapter 2. Fossil and Fresh Water Challenges

To the thirsty I will give water without price from the fountain. **— Revelation 21:6**

A solution to the challenge of water, a critical issue throughout the ages for food production and human habitation, remains possibly the most vital global issue today. Solutions to providing sufficient freshwater offer only two alternatives:

1. Find, harvest and transport more water.

2. Develop food and energy sources that require no or minimal fresh water.

The first solution replicates the unsustainable actions of the last 50 years – using wider pipes, larger pumps and deeper holes to mine more non-renewable water faster to grow crops. Even if energy were free, increased pumping crashes aquifers within a few decades and makes the land unsustainable for both crops and people. Powerful modern pumps draw water from deeper and deeper wells – at an unsustainable cost of both power and water. Unsustainable use of fossil water foretells severe hunger and starvation for future generations.

The second alternative leaves the aquifers in place to support people, businesses and food crops now and for future generations. The world desperately needs an abundant agriculture system that does not rely on freshwater.

Our world's water

Globally, irrigation for crop production claims about 70% of all freshwater and consumes about 80% of surface and groundwater in the U.S.[41] industry uses 20% and residential users consume about 10%. With the demand for water growing steadily in all three sectors, competition is intensifying. In this contest for water, farmers almost always lose to money – cities and industry.[42]

Our planet held far too little freshwater to support food production for expanding populations even before global warming began melting and evaporating our ice caps, glaciers, snow packs and reservoirs. Our faucets and fountains are already going dry because farmers are putting far too many straws in the ground, thus removing stored groundwater reserves. The aquifers on which food production depends are being depleted rapidly and many are likely to expire within a generation. Some aquifers will go dry in this generation and some have already crashed.

Water creates the primary limitation to food production because failing available freshwater, crops quickly become thirsty and wilt, stunt or die. Without sufficient freshwater delivered on time, crops fail and the land reverts to its natural state – which in much of the world is prairie or desert – and human populations must migrate to plentiful water.

Production and yield are directly related to water use. Insufficient applied water stresses crops and decreases yield. More irrigation has doubled food production over the last 30 years but at the unsustainable expense of tripling the freshwater consumed.

Much of the 300% increase in water consumption occurred because new croplands expanded into deserts. Desert regions are often productive due to the considerable solar energy but the heat consumes more water from transpiration (plant water losses) and soil moisture evaporation. Irrigation systems often lose 50% of the applied water before the water reaches the crops from leaks and evaporation.

Today, when the number of hungry people has reached record highs in America and the world, acute water scarcity has struck countries in the Middle East and North Africa, as well as Mexico, Pakistan, South Africa, the United States and large parts of China and India.[43] Iran was forced to import over a million tons of grain from the U.S. in 2007 because their crops failed due to heat and drought.

Water is incredibly expensive to move and delivery costs of $4,000 or more per acre are typically heavily subsidized by governments.[44] Surface water supplies are typically far distant from croplands and require building, operating and maintaining huge dams and reservoirs connected with hundreds of miles of canals, pipelines and pumps that deliver water. Groundwater irrigation requires pumping and as water tables drop, pumping becomes increasingly more expensive. Many farmers have found the cost of pumping groundwater to exceed the value of their crops and have had to abandon their land.

About 60% of the world's food supply consumes groundwater which poses a serious threat to aquifer depletion. The two biggest water consumers, China and India, have the largest populations. Nearly 80% of China's grain production depends on irrigated land and about 60% in India.[45] More than half of this irrigation water comes from groundwater aquifers which are being emptied at up to 300 times nature's replacement rate.

The U.S. faces severe freshwater shortages too. U.S. Geological Survey shows California's San Joaquin Valley has lost 60 million acre-feet of groundwater since 1961. Groundwater pumping continues to cause the valley floor to sink, a problem known as subsidence. This threatens the stability of surface structures such as the California Aqueduct, which delivers drinking water to more than 20 million people. About 20 percent of groundwater pumped in America comes from under the Central Valley.[46]

In normal years, California produces more than half of the vegetables and fruit consumed in the U.S. The state irrigates 9.6 million acres, using roughly 34 million acre-feet of water from lakes, reservoirs and rivers or pumped from groundwater. Lester Snow, director of the California Department of Water Resources, said in 2009 that

California may face its worst drought in recorded history.[47] The Central Valley Authority that distributes irrigation water through the heartland of California announced a zero allocation to many crop regions. The Bureau of Reclamation estimated that one million acres would be put out of production and another two million acres would grow less food than normal. Lester Snow called the situation grim because the more than half the snowpack that usually stores California's water has melted in the heat.[48]

California agriculture consumes 81% of the water in the state while providing only 3% of the state's revenue. Agriculture fights constantly with cities such as San Francisco, Los Angeles and San Diego for access to water that is diverted mostly from the Sierra Nevada Mountains in the northeastern part of the state to the fertile deserts of the Central Valley and southern California.

Drought also plagues the West, South and Eastern U.S. In the fall of 2008, Texas farmers received no fall rain and lost their winter wheat crop. Farmers will not be able to plant spring crops because there is no soil moisture. Texas farmers rediscovered the "WW problem" that farmers have known eons; "If there is insufficient Water to germinate Weeds, crops don't grow either."

Rational government policy would limit irrigation to sustainable yields from surface sources and groundwater held in aquifers. Instead, government policies in the U.S. and globally have encouraged maximizing short-term food production by subsidizing water including transport, delivery and the energy needed for pumping, Figure 2.1. When a commodity has a near- zero cost, users waste it. Inefficient and over irrigation wastes trillions of gallons a freshwater each year. Over-pumping at several times the sustainable yield has resulted in plunging water tables on every food growing continent. Many aquifers are falling at 10 feet or more each year.

Figure 2.1. Expansion of irrigation on croplands

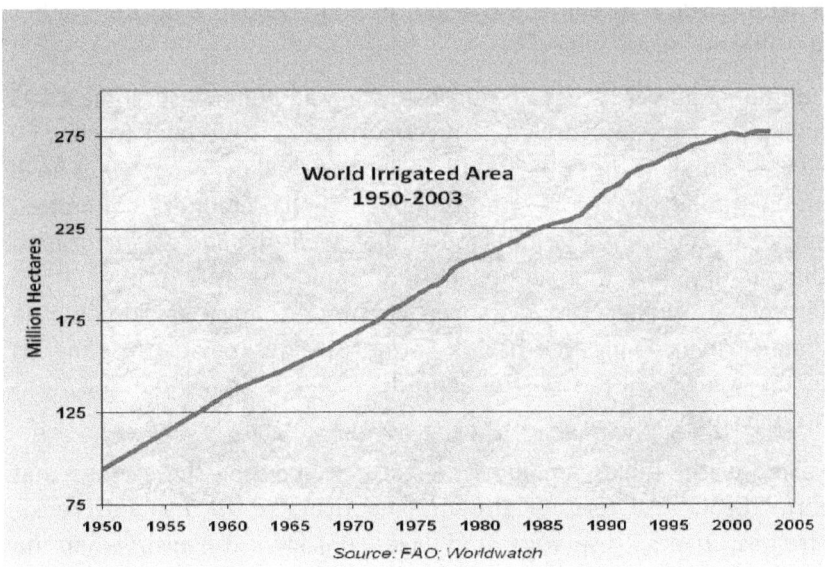

Source: FAO; Worldwatch

Water and energy

Water and energy are linked because as water tables fall, the energy required for pumping rises. As the water gets deeper, pumps eventually fail and water must be piped in at huge energy cost or the land abandoned. In some cases, farmers can practice dry land farming where they must pray for rain and know that their fields may have low productivity, even in normal rain years. Unfortunately, with the onset of global warming, normal rain patterns are historical, not present.

India and China are opening a coal-fired power plant every six days to meet the rising electricity demand from irrigation. However, the new power plants cannot keep up with the need for more energy to pump and move water. Expanding food production into deserts creates a vicious cycle from the need for constantly more freshwater and additional energy to pump or move water. Electricity blackouts are common in India and parts of China where over half of the electricity is used to pump water from depths of two thirds of a mile.[49] Wells in

China approach a mile deep and only huge government subsidies make the water available. However, extremely deep wells are neither sensible nor sustainable.

Additional power serves no purpose when aquifers crash. India's 100 million farmers have drilled 21 million irrigation wells over the last 30 years and half of those wells and millions of shallower tube wells have already gone dry. More than 25,000 of India's farmers committed suicide in 2003 when farmers found their wells dry and could not feed their family, pay their debts or find employment.[50] Another 1,500 farmers in similar circumstances committed mass suicide in one Indian province in March 2009.[51] Farmer suicides may be the canaries in the mine as industrial food production fails in various locations around the world.

Falling water tables amplify water scarcity causing the springs that form the headwaters for streams and rivers to go dry and creeks, streams, rivers, reservoirs and lakes quickly extinguish when no upstream flow feeds them. Many water sources that have been dependable for hundreds of years have gone dry and cannot support food production. The Gangotri Glacier in the Himalayas has provided 70% of the water for the Ganges River for centuries but is retreating 35 yards each year.[52] When the glacier melts, the Ganges will become a seasonal river depending on annual rains depriving 40% of India's irrigated cropland nearly half a billion people of water.

Climate change changes rainfall's frequency, intensity and duration. Monsoons, hurricanes and typhoons will be heavier, more variable and bring greater risk of flooding. Climate models show an increased incidence of drought, which threatens both crop production and livestock grazing. Prolonged droughts will raise the incidence and intensity of natural wildfires.[53] Nature's act that follows droughts and wildfires – flooding – removes irreplaceable topsoil that leaves farmers with fallow fields.

In the U.S., over half the crops by value are grown on land with insufficient rain that requires irrigation. Nearly all croplands from 100 miles west of the Mississippi River, (the 100th Meridian), all the way to the Pacific Ocean, need irrigation. America's croplands are extremely productive when irrigation provides sufficient water because they receive considerable solar energy that enhances photosynthesis. Extra sunshine enhances crop productivity but significantly increases water demands. The extra heat combined with dry winds depletes soil moisture quickly and creates the need for substantially higher irrigation volume and frequency.

The freshwater challenge becomes more complex from the invasion of salt from the accumulation of soil minerals dissolved and carried in irrigation and sea salt deposited from storm surges.

Salt invasion

Salt invasion hits plants from above and below. Food crops cannot tolerate salt because the large salt ions clog their roots and cuts off the flow of nutrients. When irrigation water evaporates, the salts are left on top of the soil. Growing crops also separates irrigation water from the dissolved salts it carries, concentrating the salts in topsoil.

When too much water is used to irrigate fields the extra water may raise the water table creating waterlogged soils. Waterlogged soils may simply drown the plants or elevate soluble salts to the root zone where the salt kills the plants.

The same destructive salts move with both surface and groundwater. As water flows downstream, human, agricultural and industrial pollutants accumulate and make the water unfit for human consumption or for irrigation. Besides pollution from chemical discharges, water carries a variety of dissolved salts. The lower Colorado River carries 2,000 pounds of salt per acre-foot and the San Joaquin River in California carries 3,300 pounds per acre-foot.[54] Every mile downstream salinity increases and, due to their heavy load of salts, neither river can be used for irrigation far downstream.

Economy of Pakistan is severely threatened by irrigation salt invasion from the Indus River which flows into the Arabian Sea. The Indus provides water for 90% of Pakistan's irrigation and half of its electricity.[55] Aggressive irrigation has created waterlogged soils in many areas, carrying 24 million tons of salt to the Sindh and Punjab plains, adding half a ton a year to every year acre. Farmers have had to abandon over 10% of salt encrusted cropland and food production has declined rapidly on the remaining fields. In the Sindh province, more than half of the white, salt crusted land is now barren. The only way to recover those precious croplands would involve leaching, which would mean heavier irrigation. However, leaching will not work because the underlying caliche (hard clay) causes waterlogged soils. Even if water logging were not a pervasive problem, additional water is not available for irrigation.

Sea salt invasion from storm surges also destroys croplands and is a function of man-made global warming that is causing oceans to rise for several reasons:

1. Heat raises water temperature causing thermal expansion and higher water levels.

2. Warmer surface temperatures fuel fierce storms that create a deluge that runs off rather than penetrating the soil.

3. Fiercer storms cause higher storm surges that carry sea salt far inland.

4. Hotter temperatures melt glaciers and arctic ice sheets, adding substantially more water to the oceans.

When the Greenland ice sheet melts, sea level will rise 23 feet. When the West Antarctic Ice Sheet breaks up, it will add another 15 feet, for a total of 38 feet.[56]

Much of the world's best cropland lies on river deltas or coastal lowlands. These lands will be ruined from sea salt invasion. In 2005, a massive, 15 foot dome of flood water inundated much of southwest coastal Louisiana before, during and after landfall of Hurricane Rita. Several counties were left with residual saltwater for several months and much of the area lays barren today due to heavy salt.[57]

Coastal communities face significant losses in fresh water as saltwater intrudes inland and populations will be forced to migrate. Recent studies show that salt water penetrates 50% further inland underground than it does above ground, ruining residential, industrial and irrigation water sources.[58] About 40% of world population lives less than 40 miles from the shoreline which will put additional pressure on croplands as they are forced to move from storm surges and salt invasion.

Consumptive water

Non-consumptive water may be consumed, recovered, cleaned and reused. City water for household use indoors is non-consumptive because water used in the home goes down the drain and empties through the sewers to the local wastewater treatment plant. The plant removes impurities, possibly with algal ponds, and dumps the clean water into a nearby stream or water source. Mississippi river water, similar to the waters of other rivers, is reused many times by cities on its way to the Gulf.

Consumptive use extracts surface or ground water that cannot be recovered and reused locally.[59] Many people think that the natural water cycle returns water that has been used. Unfortunately, the water cycle returns only a portion of the water used and not in a dependable manner.

Water used outside homes such as on lawns, gardens and in pools as well as city parks and golf courses is consumptive and lost for reuse. Household water for farms or communities that use septic tanks rather than sewers is also consumptive.

Irrigation is consumptive because the water evaporates from the soil or transpires from crops (evapotranspiration) into the atmosphere creating water vapor. Rising air currents carry the water vapor upward, high into the atmosphere, where the air cools and loses its capacity to support the moisture. The water vapor condenses to form cloud droplets, which may eventually combine with other droplets and produce precipitation. Water vapor arising from irrigation may fall as rain 2,000 miles away but is lost for local consumption. The extracted moisture falls not on cropland but floats on the wind to fall on distant mountains or oceans. From a farmer's point of view, consumptive water is gone. Lost water does not renew the local area's soil moisture, wetlands, wells or aquifers.

Crops inefficiently incorporate water into their biomass. Plants draw water through their roots to deliver nutrients to the trunk and leaves. The plant then releases most of the water through small pores on the undersides of the leaves called stomata. Crops protect themselves from heat with transpiration that acts similarly to evaporative cooling, stabilizing the plant's temperature.

To get sufficient water, farmers must apply far more water than crops actually incorporates in their biomass. Most of the excess water is lost from evapotranspiration. Some irrigation water seeps down below the plant's root zone and also is not available for reuse. Unfortunately, roughly 80% of extracted water in the U.S. goes for consumptive use in agriculture.[60]

Corn provides an example of a high water consumptive crop because its tall stature, often nine feet, provides lots of surface area for transpiration. A single acre of corn gives off about 4,000 gallons of water each day from evapotranspiration.[61] People who live near a corn field can feel the extra humidity produced by the escaping water vapor. Water vapor is a greenhouse gas that accelerates global warming as it absorbs and radiates the sun's rays. The high water loss from corn also means more consumptive use water must be provided to sustain the corn field's growth.

Aquifer depletion

More than 65% of U.S. irrigation extracts water from underground aquifers which are composed of sand, gravel and other materials with gaps large enough to hold and transmit water.[62] Aquifers display all the variability associated with surface geologic formations which include porosity, permeability and different types of rock, sand or clay at different depths. Water's highest point in an aquifer represents the water table which may be close to the surface or deep underground.

Aquifers close to the surface, called alluvial aquifers, may be partially recharged by annual rains. However, many alluvial aquifers capture and store only 10 – 30% of annual rains as the rest runs off before filtering down to the aquifer.

Large corn producing states – California, Nebraska, Kansas, Texas, Arkansas, and Idaho – account for 53% of total U.S. irrigated acreage.[63] Much of their irrigation water depends on fossil aquifers. Fossil water was trapped thousands of years ago in ancient sediments below a layer of bedrock, shale or caliche that blocks recharge.

The large Ogallala aquifer which runs from South Dakota to Texas supports the groundwater needs of eight High Plains states and runs from South Dakota to Texas. The aquifer was filled with water from a glacier melt 25,000 years ago. Even if the Ogallala were not blocked from recharge by a layer of shale, the High Plains would still be pumping water unsustainably because the region gets less than a third of the water extracted (five trillion gallons) in normal rainfall years.

Mining fossil water works similarly to oil extraction – when the pool goes dry, no more groundwater is available. Mining water for irrigation or for use by cities or industry causes the fossil aquifer water level to drop. Then progressively larger pipes and stronger pumps are necessary to extract water from deeper and deeper wells. When much of the fossil water that has been in the ground for millennia is extracted, the aquifer crashes.

Considerable variation occurs in aquifer death due to local geology. Some aquifers crash with 30% of the water still remaining, because the water cannot be extracted due to turbidity, cave-ins, pebbles or thick mud. In other cases, wells must be sunk so deep that the cost of pumping the water exceeds the value of potential crops. Many farmers on the southern end of the Ogallala aquifer in New Mexico, Texas and Oklahoma have already watched their wells go dry and their precious croplands have returned to dry prairie. Failing available water, rural families have to move and leave near valueless land to nature.

The Oglala Lakota tribe in the Badlands of South Dakota lent their name to the Ogallala aquifer, but they can no longer use its water because it has become contaminated. The waste streams for farming and sewage have ruined their groundwater and now they must haul water from the Missouri River 200 miles away.[64]

America's water crisis

Without available groundwater for irrigation, neither food nor biofuel production is sustainable. Cities are not sustainable without access to quality water, no matter how it is obtained – from groundwater or transported water. Transporting water from new sources in the U.S. is nearly impossible because all freshwater water sources have been locked up in water rights contracts for over 100 years. Cities cannot take upstream water because, in many cases, 150% of the available water is already promised by contract. Water contracts were written before global warming and drought diminished rains and available water.

Freshwater cannot be manufactured, even with the most advanced technology. Desalinization offers some hope, especially if cheap non-fossil sources of energy can be developed. Desalinization can provide a bridge for thirsty cities near coasts but is an order of magnitude too expensive for food production.

Due to drought and overdraft (pumping in excess of aquifer recharge), water tables have dropped and springs, rivers, lakes and reservoirs have gone dry all across the country including New England, the East Coast, the normally rainy Mid-West and the West. The spring that has fed Tennessee's Jack Daniels distillery for more than 100 years in Tennessee is under threat of going dry. The West is locked in a 500-year drought, leading to severe wildfires and land destruction. Fossil aquifers that are not recharged by rain have been significantly reduced by overdraft across the nation. Numerous aquifers have crashed and the U.S. Geological Survey predicts more imminent aquifer crashes, even normally rainy regions such as Arkansas.

Water levels have declined 200 feet in some areas on the Ogallala, which significantly increases the energy needed to pump water. Wells that served 120 acres a decade ago now serve less than half the area.[65] Many reservoirs and lakes on the High Plains, South and West are at their lowest levels on record with little chance they will regain once normal water levels.

William Ashworth documented how irrigation is unsustainable on the High Plains and describes the management of Ogallala water as a fast-declining resource where the aquifer will simply run out of water. He concludes that "Management of water on the High Plains looms as one of the most important American challenges of the 21st century."[66]

Table 2.1. America's Water Crisis.

America's Water Crisis	
Surface water	Significantly less surface water flows due to drought and global warming. Drought has dropped rainfall to record lows and heat reduced snow packs to less than 50% of normal across much of the West. Drought in the East and South has reduced river flows and dried up springs, creeks and lakes.
Overdraft	Overdraft occurs when more water is removed from underground water storage than nature replaces. Currently, 95% of America's water is being extracted at rates over natural replacement.
Fossil water extraction	Extraction of fossil water for irrigation is expanding. Fossil water is non-rechargeable underground water storage that came from melting glaciers thousands of years ago. Millions of America's prime cropland acres are mining trillions of gallons of fossil water. When fossil aquifers are depleted, they will be gone forever.
Agricultural pollution	Agriculture, especially row crops like corn, pollute water – streams, rivers, lakes, estuaries and well-water – from agricultural pollution which includes dissolved solids and salts along with fertilizers, pesticides and herbicides. Some corn-growing states have 90% of their streams, rivers and lakes unfit for human use due to agricultural

	chemical pollution. Agricultural chemicals have invaded wetlands and cause reproductive problems for birds, fish and amphibians.
Estuaries	Estuaries, the tidal areas where rivers and creeks reach the sea are particularly distressed. The EPA assessed 51% of estuaries as impaired and not fit for one or more uses such as supporting aquatic life, fish consumption or recreation.[67] The EPA largely exempts agriculture and ethanol refineries from critical policies like the Clean Water and Clean Air Acts. Over 50% of U.S. estuaries are unfit for aquatic life, fisheries or human recreation.
Water rights	Water rights are legal commitments by citizens, cities and other entities for control of precious surface and ground water. Even in normal rain years, many areas have over 100% of the local water promised under contracts. In dry years, many areas simply extract more ground water – a strategy that works until the wells go dry.

Water consumption in food

Today two out of three world citizens subsist on local or imported grains and local plants because they cannot afford meat. Livestock represent the single largest user of land, as meat production accounts for 70% of all agricultural land and 30% of the land surface of the planet.[68] In order to produce meat, roughly one-third of the world's food grains go to feed livestock. In addition, livestock are responsible for 18% of all greenhouse gases – more than all cars and SUVs combined.[69] Livestock contribute 37% of the methane and 65% of the nitrous oxide to the atmosphere which are worse greenhouse gases than CO_2. A single cow-calf pair produces more gas emissions than a person driving 8,000 miles in a mid-size car.[70]

Meat is dramatically underpriced relative to plant foods because meat production benefits from both crop and Big Oil subsidies. Meat prices reflect no environmental accounting even though meat producers extract trillions of gallons of fossil water, billions of gallons of fuels and millions of tons of chemicals. In addition, agricultural waste streams are the primary polluters of surface and groundwater, yet the public costs and resource losses are absent from meat prices.

America and the new consumers have become carnivores and demand animal-based foods, which create an ecological cost of multiple pounds of grain for each pound of meat or dairy. America has doubled its per capita meat consumption since 1990 and people around the world desire to emulate the consumption patterns of the rich.[71] The added demand for animal food substantially reduces available food grains. Example feed conversion ratios, grain to meat, are:

- Farm raised tilapia or catfish   2:1
  (much of fish weight gain is water trapped in tissues)
- Poultry – chicken or turkey   3:1
- Beef, pork or lamb   8:1[72]

Since one ton of grain consumes 1,000 tons of water, one ton of beef on the hoof uses 8,000 tons of water. However, beef on the hoof contains 57% waste – bone, hide, fat, stomach and unsellable organs. A ton of beef on the hoof yields only 43% sellable beef, so 2.3 tons of beef on the hoof are necessary to yield a ton of sellable beef.[73] Therefore, a ton of sellable beef, based on grain input, consumes 18,400 tons of water or 9.2 tons of water per pound of beef.

Actual water required by beef is even higher because feed conversion ratios reflect only the grain and ignore the cow's own water consumption. A 1200 pound cow drinks about 36 gallons of water a day.[74] The cow's thirst adds another 20% to the water cost of beef. A pound of sellable beef consumes 11 tons of water or 2,650 gallons. In spite of the huge water cost of meat, meat production continues to increase due to demand, especially in India and China Figure 2.2

41

The water cost of sellable beef translates to a water cost for a quarter-pound hamburger of 3,000 gallons plus another 400 gallons for the cheese, 100 gallons for the lettuce and tomato and another 100 gallons for the bun. A gallon of milk costs 4000 gallons and a pound of coffee also consumes about 2,650 gallons of water to produce.[75] The water cost of beef is important because the U.S. is one of the world's leading exporters of beef. The combination of meat and ethanol production will accelerate depletion of limited and non-renewable U.S. groundwater.

Figure 2.2. Global Animal Production

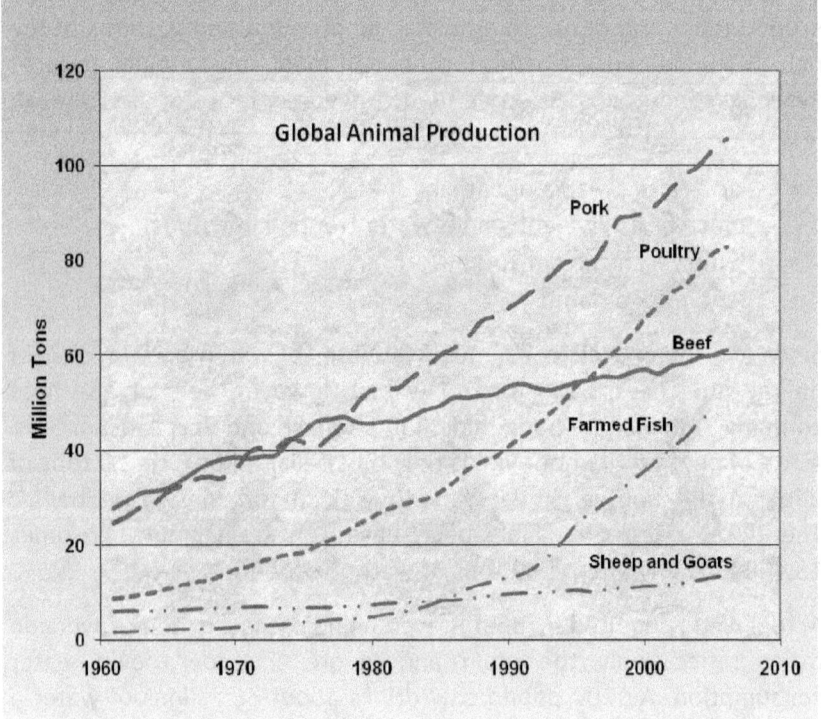

By 2010, China will have twice the number of new consumers as the U.S., 600 million people, who will demand higher value foods that create a high water exchange. China already leads the world in production and consumption of meat. Chinese per capita meat and milk consumption have doubled in the last 20 years.[76] China's increasingly high demand for grains combined with its distressed ecology and water problems (The two largest rivers, the Yellow and Juma, have gone dry, as have nearly all the lakes around Beijing.) will dramatically reduce world grain supplies. When drought, storm or pest invasion destroys food production in a major food growing region such as China, India, Russia, Ukraine, Argentina or the U.S., food markets will be destabilized worldwide.

Even with all the advances in technology and weather forecasting, humans have become more vulnerable to small climate disruptions. When past civilizations experienced drought, such as the U.S. Dust Bowl of the 1930s, people could simply migrate to another area with more water. Current populations are far denser and millions of people have nowhere to flee.

Water consumption in the U.S.

America's water scarcity is self-imposed by substantial exports in virtual water contained in meat and grains and the use of corn as a biofuel. Virtual water is the amount of water required to produce the food or product. Corn consumes about 80 million prime cropland acres each year. The U.S. is the world's dominant corn exporter, averaging 70% of world corn exports during the 1990s. About 20% of the U.S. corn crop is exported and another 27% is used for ethanol.[77]

Grain exports represent roughly 124 billion tons of virtual water. Ethanol production consumed another 100 million tons of corn that used 100 billion tons of water. The USDA reports show that meat exports in 2007 were 771,196 million tons, down from 1,279,717 million tons in 2000. Most Americans are unaware of the significant water lost in grain and meat exports because government crop, water and power subsidies are not accounted for in the price of either food grains or meat exports. Economists refer to virtual water in food crops but to farmers, the water is not virtual; it is real.

Each gallon of ethanol produced with irrigated corn consumes 3000 gallons of freshwater for the corn feedstock and additional water for refining.[78] Some areas on the high plains get about 1/3 of their water as rain so their ethanol consumes 2,000 gallons of water per gallon of ethanol. The U.S. produced nine billion gallons of ethanol in 2009 which consumed about two trillion gallons of fresh water for irrigation while displacing less than 3% of U.S. oil imports. Ethanol production using irrigated cropland is neither sustainable nor sensible.

Table 1.2 U.S. Fossil Food Exports

Fossil food	Millions of metric tons	Fossil food	Millions of metric tons
Corn	54	Soybeans	29
Wheat	34	Sorghum and barley	7

High Plains' farmers know that they will be the last generation to mine water from the fossil aquifers. When the water is gone, the land will return to prairieland that some have called a Buffalo Commons. Every farmer has heard and read countless reports on water scarcity, especially the eventual loss of critical aquifers such as the Ogallala. This knowledge sets off a response called the "Tragedy of the Commons" where farmers chase short-term wealth with full knowledge that they are destroying their collective future. No one can afford to miss out on the boom, because they will all share in the eventual bust. The result is denial of the obvious facts and overconsumption of a scarce public resource. Unfortunately, the common greed accelerates aquifer depletion.

Unsurprisingly, public policies, such as using corn ethanol as a biofuel encourages significant water waste. The cost of irrigation water depends on the water district and many farmers receive not only crop subsidies for the corn they grow but water subsidized at 2% of the residential rate. Other farmers not only receive 100% water subsidies but power subsidies to pump and transport their irrigation water. No one has calculated the public subsidy cost for irrigated corn used for ethanol but it probably exceeds $25 per gallon of ethanol. Nearly every state in the West that depends on irrigation to produce corn has one or more ethanol plants. The typical ethanol plant produces 50 million gallons of ethanol annually which consumes 150 billion gallons of irrigation water for the corn feedstock.[79]

Cities on the High Plains will soon be forced to decide if they can afford to pipe in water or if they must disburse their residents. Cities such as Aurora Colorado, Atlanta Georgia, Orme Tennessee and Palm Springs California came very close to running out of fresh water in 2008 and now realize they face very tough water decisions. Similar tough choices are being imposed on communities today in several parts of the world including China, India, Mid-East and North Africa.

America has insufficient disposable water to waste. The U.S. must end ecologically destructive subsidies for irrigation water and ethanol or millions of prime cropland acres will be dry and dusty within two decades. Growing food and biofuel crops will be only a memory because no irrigation water will be available. When aquifers crash and wells go dry, there will be no water for growing animal feed. Farmers will have to slaughter their livestock because the cost of animal feed will be too high. Farm families will be forced to migrate similar to the Dust Bowl era because their once precious land will become nearly valueless.

Nearly one half of America is desert. It may be time to rethink water policy, especially burning our water in the form of corn ethanol and exporting subsidized water in the form of corn, wheat and meat.

Sustainable water use for food

Constraints for sustainable water used in food and energy production include the following.

Table 1.3 Sustainable Water Use

Issue	Description Fossil fuels and ethanol	Needed solution for food and energy
Irrigation	America has insufficient freshwater for irrigating both food and biofuel crops.	No or minimal irrigation needed.

Fresh water	Coal and nuclear power both consume over 2 billion gallons of water for cooling. Ethanol consumes 2 trillion gallons for irrigating corn.	No or minimal fresh water needed.
Water pollution	Mining coal and corn production severely pollute water.	No water pollution
Air pollution	Severe air pollution from mining and growing corn and from burning gas and ethanol.	No or minimal air pollution from production or use.
Ecologically sensitive	Mining and corn production are ecological disasters due to erosion and wetlands destruction and strip mining.	No or minimal pollution of soils and ecosystems.
Affordable	Expensive production subsidized by government.	Cheap production using plentiful inputs.
Secure	Long, insecure supply chain and high transportation cost.	Short supply chain and low transportation cost.
Sustainable	Mining ag chemicals and massive corn production are not sustainable due to their high consumption of fossil resources, especially water.	Leaves sufficient water for the next five generations.
Efficient	Significant energy waste in production.	Minimal energy waste in production.

The *Koran* offers useful advice from a source that knows the value of water. Water in a desert is the source of all real property.

- No man may abuse a well.
- If an owner has a surplus of water, he must give some to strangers and their cattle.
- He has no obligation to provide water for irrigating crops.

Many cities will be citing the Koran in the next few years as their citizens struggle for water security. The challenges associated with sustainable and affordable food are critically important because growing food requires considerable freshwater and energy to pump and deliver the water.

SAFE production

Sustainable and affordable food and energy, SAFE production saves freshwater and produces food and energy in wastewater, brine or salt water and cleans the water sufficiently for reuse in irrigation. Israel currently cleans and recycles 87% of their municipal, industrial and agricultural waste water with green solar energy – algae.[80] In closed and semi-closed production systems, green solar energy recovers, recycles and reuses nutrients from wastewater while producing food, fodder and fertilizers in a manner in an ecologically positive manner.

Chapter 3. Fertile Topsoil

The threat of nuclear weapons and man's ability to destroy the environment are really alarming. And yet there are other almost imperceptible changes — I am thinking of the exhaustion of our natural resources and especially of soil erosion — and these are perhaps more dangerous still, because once we begin to feel their repercussions, it will be too late.

The Dalai Lama's Little Book of Inner Peace[81]

Soil degradation and erosion is a slow insidious process that, as the Dalai Lama warns, jeopardizes food production. More than 99% of human food comes from the land-based food crops. Less than 1% comes from aquatic ecosystems, and industrial fishing fleets have reported diminished catches each year for the last decade.[82] The world's food supply depends upon sustaining fertile soils to produce food. In order to maximize short-term production, industrial agriculture manipulates the natural ecosystem and disrupts, degrades, displaces and destroys precious soils and fossil nutrients. Around 33% of cropland the world's cropland has been abandoned in the last 30 years due to soil wear-out or erosion.

Topsoil

Plants depend on fertile topsoil from which they draw the nutrients that support photosynthesis, cellular vitality, energy transfer, structure and fruiting, which produces seeds. When plants have insufficient topsoil, or the eroded soil does hold soil moisture, plants may germinate and even grow partially but fail to set their fruit. Plants without seeds cannot propagate and do not provide food because animals and people typically eat the "fruit of the vine."

A single inch of fertile topsoil may take nature 500 years to create.[83] Many crop producing regions have thin topsoil only 4 – 6 inches deep and when the topsoil degrades or erodes, food production stops. Topsoil must be loose enough to enable water to penetrate because plant roots depend on soil moisture to dissolve nutrients. Only dissolved elements and compounds can be absorbed and transported through the plant.

Topsoil forms a tapestry of loam – sand, silt and clay – and organic humus that must allow water to infiltrate but not percolate through the root zone too fast. Topsoil acts like an Oreo cookie in that the loam (dirt) provides physical support for the plant's roots and anchors the plant while the humus provides delicious and valued nutrients. Below the roots, rock and clay hold the humus in place.

Humus stores about 95% of the organic nitrogen and phosphorous that plant's need.[84] Humus gets its dark color from the rich organic carbon it stores, aids in water retention, and hosts diverse colonies of microorganisms which breakdown organic compounds into plant usable nutrients. Plants absorb dissolved nutrients through their roots and depend on the retained water in the soil to transport nutrients. Food crops cannot grow effectively when topsoil has been eroded due to problems with water retention – too much or too little – and lack of organic carbon and microorganisms.

Growing food disrupts the natural soil biodiversity created over eons. In one year, cultivation may erode the topsoil nature took a century to make. Cultivation removes weeds but breaks up the soil aggregate which leads the earth vulnerable to erosion from winds, rain and irrigation. Cultivation so depletes soil nutrients, pollutes the soil, air and water that some farmers have begin to practice "no till" agriculture.

Farmers' ability to grow crops year after year depends on a combination of crop rotation and substantial added agricultural chemicals, especially nitrogen, potassium and phosphorous. However, when the topsoil erodes, even chemicals cannot restore soil fertility.

Fertile Topsoil

Paul Ehrlich, possibly the most knowledgeable scientist in the world on the environment, notes:

> Of all human activities, agriculture arguably has the greatest environmental impact, especially the destruction of biodiversity – plants, animals and microbes that share Earth with us and upon which our lives depend. … Agriculture itself can destabilize the very process it depends upon for success.[85]

About 40% of the world's remaining agricultural land is seriously degraded. The worst affected regions are Central America, where 75% of land is infertile, Africa, where a fifth of soil is degraded and Asia, where 11% is unsuitable for farming.[86] According to the U.N., an area of fertile soil the size of Arizona plus New Mexico is lost every year because of deforestation, climate change and erosion.

Erosion

The key to soil fertility and the impact of erosion centers on the valuable organic carbon and microorganisms, stored largely in humus. When organics and microorganisms are present, plants flourish. For thousands of years, farmers harvested the food grains and left the stover (residue) to enrich the soil's organic matter and enhance the humus. Modern farmers tend to harvest the entire plant leaving no replacement organic material, using synthetic fertilizers and mined chemicals as a substitute for the soil organics.

Topsoil Erosion

Soil erosion depletes topsoil fast and furiously and without equity. Winds and water take 5:1 the rich organic matter compared to loam because humus tends to be lighter and often flakes in the process of organic breakdown.[87] (Think of the organic breakdown of a leaf.) Wind tends to kite the organic matter and blow it into dust clouds while water simply floats it away. Both actions accelerate erosion of the most valued soil component first -- humus. Humus loss causes clear-cut rainforest land to degrade after only two growing seasons. Rainforest land is often hilly and rocky with thin topsoil which erodes easily. Degraded soil may retain some vital plant nutrients but without water retention, the nutrients do not dissolve and are superfluous because they are unavailable to the plant.

Rain contributes to erosion. Raindrops appear to be innocent. A raindrop falling in still air arrives at about 20 mph. Storms that drive rain propel each raindrop at the speed of the wind, possibly 60 mph, at an angle. Millions of raindrops from a single storm can send soils plunging downhill and into creeks, rivers and lakes. The 2008 floods in the Midwest left washouts 30 feet deep which removed not only all the fossil topsoil but all the applied agricultural chemical pollutants, depositing them in waterways.

Raindrops smash into topsoil and break apart the soil aggregate. The individual sand, silt and clay particles change from solid to liquid and the particles fill the soil pores and reduce infiltration – water penetration. When surface pores become clogged with sand, silt or clay, overflow of water – runoff – occurs.

Farmers cultivate their fields to clear weeds and loosen soils to enable improved water penetration. Tilling cropland adds to fossil fuel costs and accelerates soil erosion because it breaks apart the soil aggregate and plant roots which hold the aggregate together. Cultivation breaks the topsoil into fine particles which provide a great seedbed for plant germination but leave the soil highly vulnerable to the capricious actions of winds and water. The U.S. Department of Agriculture, (USDA) estimates average annual soil erosion from corn-producing land approximates six tons per acre or about 0.05 inches.[88] Each cropland acre loses about 54 pounds of nitrogen, 13 pounds of phosphorus, 264 pounds of potassium and 132 pounds of calcium annually which is typically replaced with mined chemicals.[89]

The loss of topsoil the depth of a fingernail may seem trivial, but it is important for several reasons:

1. Topsoil is only a few inches deep in many areas.

2. The estimates are rough and erosion variability extremely high — a single storm can take an inch of topsoil from a field, far more if the field slopes.

3. Global warming brings more hot dry winds that extract soil moisture and generate huge dust clouds.

4. At six tons an acre, annual corn plantings contribute about 500 million tons of tons of silt clogging and filling waterways and all the embedded agricultural chemicals polluting the water.

Soil erosion plagued Kansas during the dry and windy winter of 1995 when storms carried away 23 tons of fossil topsoil per acre, more than two inches.[90] Soil erosion costs the U.S. about $37.6 billion each year.[91] The Iowa Environmental Resources Council estimates that Iowa, the largest corn producing state has lost about half its topsoil and continues to lose topsoil at approximately 30 times nature's replacement rate.[92]

The amount of topsoil transported downstream by the Mississippi River at Winona is about 302,000 tons per year, or 827 tons per day. The fierce 2008 Midwest floods probably carried 20 times the normal soil load. The silt carries tons of fertilizer, pesticides and herbicides into the fragile Gulf of Mexico ecosystem creating a dead zone. The dead zone at the mouth of the Mississippi River was larger in 2009 than the state of New Jersey.

Water causes about 60% of cropland erosion, while wind contributes 40%. Soil erosion displays high variability based on soil type and depth, retained moisture, slope, land management such as contour plowing, crop rotation and no-till farming. Drought may accelerate the rate of erosion, especially if strong winds, rains or storms occur.

Fierce storms

Global warming causes less rain to fall. The period between rains tends to be longer and rain falls not as farmers' rain (a slow drizzle that soaks into the soil) but in violent storms. Downpours and floods do treble ecological damage as they erode thin topsoil, runs off rather than leaching dissolved salts and moves fertilizers and agricultural chemicals into waterways. Global warming brings more violent storms and ruinous floods every decade, Figure 3.1.[93] The Union of Concerned Scientists predicts that 100 year floods are likely to occur once a decade.[94]

Figure 3.1 Floods Rising in Recent Decades

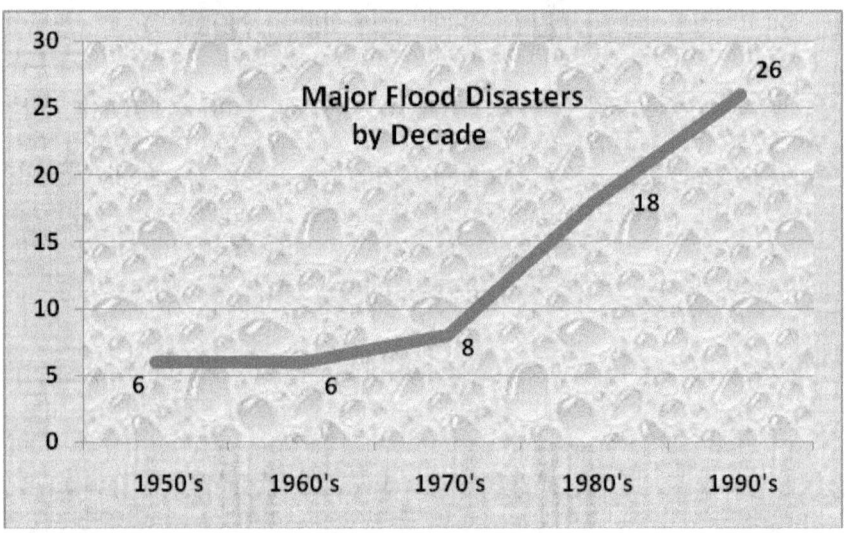

Row crops such as corn increase soil erosion from both wind and water; there is nothing to slow the wind or water as it blows or flows down the rows. Soil erosion typically occurs at the top of the soil where fertilizers, pesticides and herbicides are applied which causes pollution to the air and creeks, streams, wetlands and lakes. A 2007 Iowa Department of Natural Resources report indicated 274 Iowa waterways were seriously polluted. Fertilizer run-off causes such a problem that Iowa boasts the largest and most expensive nitrate removal plant in the world.[95]

About 78% of the corn planted in the U.S. uses genetically modified (transgenic) seeds which also contribute to erosion because the seeds need extensive cultivation.[96] Transgenic crops produce higher yields but cannot compete with natural grasses – weeds. Proof of their inability to compete with natural plants can be seen in the second year when a field swarms with weeds rather than volunteers of the transgenic crop. Consequently, farmers must till the soil before planting to remove weeds that would compete for soil moisture and nutrients. Transgenic corn typically receives additional cultivation and herbicides to control competing weeds while the crop is growing.

Land stewardship also makes a substantial difference in soil erosion. The majority of U.S. farmers rent cropland and they allow significantly higher erosion rates than farmers who own their cropland.[97] The USDA reports that core breadbasket states have farmland rentals of 65% in Iowa, 74% in Minnesota, 84% in Illinois and 86% in Indiana. Owners seeking short-term profits have far less incentive than farmers who work their land to preserve soil and water. Rental farmers fail in their responsibility as land stewards, acting instead like transient financial executives whose mantra is "Let me maximize my profit and be damned with the consequences to others, including my children."

Soil nutrients

Growing a food crop for one season removes about 50% of the soil nutrients. Without nutrient replacement with fertilizer, the next crop lacks critical nutrients and production diminishes. Farmers found that adding more fertilizer and irrigation, they could significantly increase production. However, mined (inorganic) fertilizers are inefficiently absorbed by plants so farmers must add substantially more fertilizer than the crop actually needs.

The soil nutrient problem is amplified by bioavailability, which means plants need nutrients in a form they can absorb, transport and metabolize. Fossil soil nutrients were laid down over eons and typically are immediately usable by plants. Applied fertilizers, even manure, are only available to the plant after microorganisms in the soil break down the nutrients. Full breakdown of agricultural chemicals depends on many factors and may take several years.

Soil nutrients, especially nitrogen and phosphorus, dissolve quickly in water which makes them ideal for plant absorption. The downside of the high dilution rate is that these nutrients are easily rinsed out of topsoil by rain or irrigation water. Soil degradation reduces yields before the land must be abandoned. Global grain production continues to increase but yield increases are decreasing each decade largely due to soil degradation, Figure 3.2.[98]

Figure 3.2 Global Grain Yields – Annual percent increase by decade

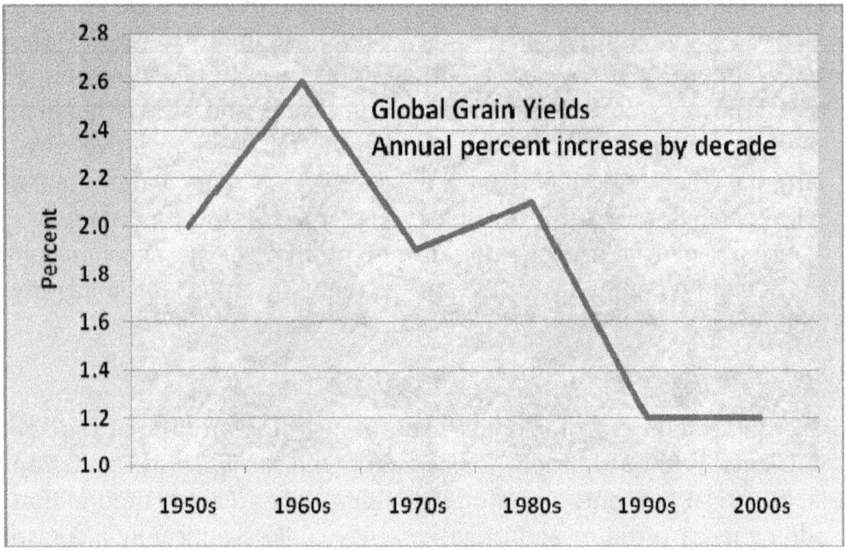

China offers an unfortunate case study on soil erosion and air pollution. Chinese farmers use extremely heavy fertilization to compensate for degraded soil. Soil organics near the surface that have absorbed the applied agricultural chemicals are carried high into the atmosphere in the spring when strong winds are common. China's air pollution caused over 20 million people to suffer respiratory illnesses in 2007. The country's health ministry demanded that the World Bank remove mortality calculations from a report on the country's air and water pollution because the numbers could trigger social unrest.[99]

Fertile Topsoil

FAO reports show that China applies agricultural chemicals more aggressively than U.S. farmers. Heavy application may improve productivity in the short term but creates two intermediate-term problems: worn-out soils and pollution. China loses over a million cropland acres a year to degradation and erosion. China's fertilizer run-off has turned lakes into algae-covered cesspools. More than half of China's waterways are so polluted that fish are dying and water is unsafe for drinking or irrigation. The Yellow River carries 1.6 billion tons of sediment into the ocean each year that has created a huge dead zone in the South China Sea that is devoid of marine life.

Water pollution from agriculture has become so bad that 300 million people – almost one-quarter of the population – lack access to clean drinking water.[100] Cropland sediment reduces lakes and reservoirs capacity and creates extra weight on the back of dams. The official Xinhua news agency admits that more than one-third of the country's 85,000 reservoirs and dams have "serious" structural problems.[101]

Soil fertility combined with climate change has slowed or stopped the rate of growth in food productivity in many parts the world. A new form of food production that does not depend on soil fertility and is largely unaffected by climate change would be a great benefit to mankind.

SAFE production

SAFE production preserves fertile croplands by growing food and energy with water-based plants that can be sited on wasteland, deserts, rooftops, parking lots or oceans.

Chapter 4. Fossil Fuels

The fossil fuel used in 1997 consumed 422 years of all plant matter that grew on the entire surface and in all the oceans of the ancient Earth about 400 million years ago.

– CBC Archives[102]

Those who debate when oil reserves peak may miss a far more perilous concern – net zero oil exports. Net zero exports occur when exporting countries decide to use the oil themselves rather than sell it on the open market. As oil reserves shrink, oil exports are likely to decline sharply because they have more value in-country rather than on the world market. Net zero oil exports means there is no oil to buy.

The problem for fossil fuel consumers occurs because the cash infusion from oil exports in Venezuela, for example, stimulates domestic consumption of government subsidized 19 cent a gallon gasoline. As reserves fall, oil prices rise, bringing in more cash and further increasing domestic consumption, to the detriment of exports.

Geologists Jeffrey Brown and Samuel Foucher built economic models for net exports that show once oil production in an exporting country peaks and begins decline, exports drop precipitously.[103] Due to increased domestic demand, only about 10% of post-peak oil production is exported. Why should a country export crude oil when they can use the oil domestically to create products such as plastics that are worth 5 to 10 times more than crude oil? Their most likely case scenario predicts that the top five oil producers will approach net zero exports around 2031.

Another petroleum engineer, Jean Laherrère, assumed greater Saudi oil reserves and projected net zero exports by 2050.[104] Unfortunately, a decade before net zero exports occur from Saudi Arabia, fossil fuels will be too expensive for industrial food production. The net zero export problem makes finding a food supply free of fossil fuels both mission-critical and urgent for the survival of human societies.

Fossil fuel challenges extend far beyond concerns about supply. They include subsidies, consumption, pollution, health impacts, prices and supply chain. Each must be significantly reduced to avoid catastrophic outcomes. The best solution; develop a suite of green energy sources that are sustainable and non-polluting and at the same time design ways to significantly reduce energy consumption. Energy savings needs to begin with farmers because, after cars, agriculture consumes the most fossil energy in the U.S., about 20%. Farms are also responsible for about 37% of America's air pollution and a majority of soil and water pollution.

Fossil foods

Farmers exhibit fossil exuberance because, thanks to government shelters and subsidies, they pay only about 20% of the true cost of their fossil inputs. In addition, farmers are sheltered from paying for externalities (social and environmental costs) from:

- Pollution of air, soils and groundwater.
- Loss of non-replaceable fossil water.
- Consumption of agricultural chemicals.
- Cost of military protection of oil assets.
- Health costs associated with agricultural pollution.

Farmers pay little for their contribution to climate change, they do not pay for the health impacts from agricultural pollution and they do not reimburse fishermen for the billions in lost revenue due to the dead zones caused by agricultural runoff.

Subsidies and cost calculations reflect neither future costs nor resource loss. As fossil resource supplies such as water and agricultural chemicals diminish, they will increase in cost. Ten years from now, farmers pumping water from fossil aquifers will pay more because they will have to use more energy and larger pipes to pump from deeper wells. The children of today's farmers will likely face the end of resource reserves and have to give up farming. Their family farm will lose most its value. If those young people were enabled to put a price on the loss of fossil resources, they would defeat subsidies for Big Oil and ethanol.

The U.S. government subsidizes fossil agriculture on the order of $12 per gallon of diesel fuel.[105] Farmers may pay $3 per gallon but the price reflects only the direct cost of the fossil fuel supply chain to the farm. Farmers benefit from huge tax subsidies to the oil industry such as the oil depletion allowance. Terry Tamminen calculated in *Lives per Gallon*, that the true cost of fuel is $15 per gallon.[106] These subsidies are built into all agricultural inputs including water, power, equipment and agricultural chemicals.

Fossil biofuels

Reasonable arguments may be made for subsidizing the domestic food supply but not food-based biofuels. Subsidies for corn ethanol, in addition to all the crop subsidies already in place, is not only foolish policy but antithetical to American values.[107] Farmers will produce the target 9 billion gallons of ethanol in 2009, which will consume 40 million prime cropland acres, two trillion gallons of freshwater, five billion gallons of fossil diesel fuel and millions of pounds of agricultural chemicals. The 100 tons of corn used for ethanol production will consume non-renewable resources, pollute air, soils and groundwater, create health problems for people and animals while replacing less than 3% of U.S. oil imports.[108] In addition, removing food from the market will drive food prices up.

The U.N. published its first report on bioenergy in 2007 and repeated warnings made by scientists, economists, food policy experts and environmental groups:

> Using food crops for biofuels threatens to destabilize world food production.[109]

The report says that burning food for fuel creates:

- A threat to food security by eliminating large quantities of food from the world market.
- An increase in food prices – beyond the means of many countries and their people,
- Serious environmental problems.
- Conversion of more food producing land to biofuel production,
- Over-demand and over-use of the best cropland.
- Over-use of scarce water resources.
- Monocropping just one crop, corn, which leads to significant biodiversity loss, soil erosion and nutrient leaching.
- Overuse of environment-damaging chemical fertilizers.
- Environmental damage from herbicide and pesticide run-off.
- Concentration of energy and food production in the hands of a few
- A substantial increase in poverty.[110]

The U.N. makes clear that a singular focus on biofuels without consideration of the unintended consequences on people and the environment puts the U.S. on a slippery slope. People, hunger and food stability need to be considered in concert with energy requirements.

Fossil fuels – coal, oil and natural gas – currently provide more than 85% of all the energy consumed in the U.S., nearly two-thirds of the electricity and virtually all of transportation fuels.[111] Electric power plants in the U.S. pollute the atmosphere each year with more than two billion tons of carbon dioxide, 14 million tons of sulfur oxides and 9 million tons of nitrogen oxides. Electric power plants waste about two thirds of the heat produced from burning coal to produce electricity due to inefficiencies in heating water to power electric turbines. Nearly all the waste heat is released into the atmosphere or absorbed by cooling water, which is then discharged to the ecosystem.

Carbon particulate pollution (black soot) occurs from incomplete combustion of fossil fuels in sources such as planes, trains, ships, trucks, residential cooking and heating fires burning solid fuels such as coal, wood, dung and agricultural residues. Black soot warms the atmosphere by absorbing solar radiation and storing the heat. It also darkens the surface of ice and snow reducing the albedo – the ability to reflect sun light – which increases absorbed heat and accelerates melting. Each tiny black particulate acts like a hot rock, absorbing solar energy and melting through the ice or snow.

Globally fossil fuels are changing the weather and climate and putting human societies in peril. Fossil fuels continually add more heat-trapping gases to the atmosphere. Burning fossil fuels account for about 75% of annual CO_2 emissions from human activities. Deforestation—the cutting and burning of forests that trap and store carbon—account for another 20%.

Coal

Farming and food processing consume tremendous amounts of electricity generated by coal. The U.S. has 600 coal fired power plants that supply 60% of America's electricity. Burning coal is the leading cause of smog, acid rain, global warming and air toxins. Each year, a 500-megawatt coal plant generates millions of tons of pollution.

Table 4.1 Coal-Fired Power Plant Pollution

500 Megawatt One Coal-Fired Power Plant Pollution per Year	
CO2 – 3.7 million tons	Carbon dioxide accelerates global warming.
Sulfur – 10,000 tons	Sulfur causes acid rain that damages forests, lakes and buildings. Sulfur dioxide forms tiny airborne particles that can penetrate deep into lungs.
Black soot particulates – 500 tons	Small black soot particulates which cause chronic bronchitis, asthma and other respiratory disease and premature death. Particulates cause haze, obstruct visibility,

	reduce sunlight and diminish crop yields.
Nitric oxides – **10,200 tons**	Nitric oxides, NOx, form ozone or smog and inflame the lungs. Nitric oxide burns through lung tissue making it difficult for many people to breath and creating higher susceptibility to respiratory illness.
Carbon monoxide – **720 tons**	Causes headaches and places stress on people with high blood pressure or heart disease.
Hydrocarbons – 220 tons	Hydrocarbons and other volatile organic compounds form ozone and cause haze and smog.
Mercury – **170 pounds**	Only 1/70th of a teaspoon deposited on a 25 acre lake make the fish unsafe to eat because fish concentrate mercury.
Arsenic – 225 pounds	Arsenic causes cancer in one out of 100 people who drink water containing 50 parts arsenic per billion.
Heavy metals	114 pounds of lead, 4 pounds of cadmium, other toxic heavy metals and trace amounts of uranium[112]
Coal plant waste	125,000 tons of ash and 200,000 tons of sludge from the smokestack scrubber each year. In the U.S., roughly 75% of this waste is dumped in unlined, unmonitored landfills.
Toxic waste	Arsenic, mercury, chromium, and cadmium contaminate drinking water supplies and damage vital human organs and the nervous system. People who drink groundwater contaminated with arsenic from coal power

	plant wastes are at high risk of developing cancer.
Cooling water – 2 billion gallons	Over 2 billion gallons of water is used for cooling. The non-evaporated water that may be 25° F hotter when released back into the environment which causes thermal pollution and decreases fertility and vitality in fish and other water creatures. Power plant water intakes suck in millions of fish eggs, fish and other water creatures. Power plants add chlorine or other toxic chemicals to their cooling water which are discharged back into the environment.
Coal transport – 1 million tons of NOx & 52,000 tons of particulates	About 14,600 railroad car loads supply 1.4 million tons of coal a year. Train engines run on fossil diesel and emit tons of CO_2.[113] Fossil diesel in locomotives also releases nitric oxides and black soot. Storing coal also releases significant black soot and tiny particulates.[114]

Both surface and underground mines creates havoc on the ecology. Mines use millions of gallons of imported fuel oil for the excavation and transportation equipment. Mine tailings often leach heavy metals and pollute ground and well water. Coal miners have one of the riskiest jobs in the U.S. and suffer from dozens of coal-related illnesses that are shared by their families who also live in the plume of coal pollution.

The quantification of the medical impacts of fossil fuel use should also motivate the development of clean energy. A coal plume downwind of a coal fired power plant's supply chain includes the coal mine, rail transportation, coal storage at the site and the power plant exhaust plume. Poor coal miners and power plant operators are trapped in the coal plume with their families. The coal plume causes social injustice for the socially disadvantaged people who suffer health risks and diminished property values.[115]

Table 4.2 Adverse Medical Conditions from a Coal Plume

People have increased health risk similar to heavy smokers. Health comparisons are made to national averages.

- Kidney disease – 70% higher
- Lung diseases – 64% higher
- High blood pressure – 30% higher
- More hospital stays
- Higher rates of premature death

A joint study by the medical schools of Brigham Young and Harvard published the first study that demonstrated cleaner air lengthened life spans significantly. They found from comparative city air quality metrics that urban dwellers' life expectancy rose by an average of 2.72 years from 1980 to 2000, and five months of that increase was attributed to breathing cleaner air.[116]

Power industry officials and politicians have pretended that "clean coal" offers a solution. However clean coal remains an oxymoron because science has not found a way to make electrical production with coal environmentally clean at a reasonable cost.

The U.S. has sufficient supplies of coal to supply the electrical energy grid. However, the ecological pollution associated with mining, transporting and burning coal – and the human health effects – make coal an unsustainable energy source even if coal were not a major contributor to climate change.

Fossil consumption

America's 300 million people represent only 4% of the world population but Americans use 25% of the world's oil, about 21 million barrels a day. America's addiction to oil increases consumption 3% each year. When the 1973 oil crisis hit, the U.S. imported 24% of our oil. In 2008, imports approached 61%.

American's dependence on imported oil means the U.S. economy is held hostage. The U.S. paid foreign countries $400 billion for oil in 2008, and another $60 billion for energy security – not counting the war in Iraq.[117] Projections suggest the U.S. will send $10 trillion to foreign nations for oil imports in the next 10 years, ravaging the U.S. economy.

Many Americans are eager for electric cars. Unfortunately, as long as electricity drawn by plug-in cars is provided by fossil fuels, electric cars will actually create roughly double the air pollution and global warming that gasoline engines create. Each unit of coal pollutes the atmosphere with greenhouse gasses at about twice the rate as gasoline.

American consumers have been shocked by significant increases in fuel prices. Unfortunately, diminishing supply combined with rising demand will continue to drive up fuel prices, creating a rising tide of prices for fossil fuel-driven consumption including food, consumer goods, transportation, medicines and recreation.

In addition to economic chaos, fossil fuel pollution causes significant impacts to human health, vitality and lifestyle. Damaging air pollutants include sulfur dioxide, particulate matter—a mixture of extremely small particles and water droplets—ozone, and nitrogen dioxide. These pollutants cause respiratory diseases, kidney disease and cancers.

Fossil fuel risks

Global dependence on fossil fuels poses four significant risks for the survival of human societies.

Table 4.3 Survival Risks for Human Societies

Occurrence	Survival Risks
1. Reserves go dry	The depletion of fossil fuel reserves will leave people very hungry, thirsty, cold and in the dark. People will not be able to afford food, water or transportation.

2. Political instability	The geopolitical strife from competition for declining fossil resources will lead to economic disturbances, political destabilization and war.
3. Mass migration	Lack of affordable energy to pump, clean and transport water will force entire populations to migrate to diminishing freshwater sources. Future wars will be fought over water, the proxy for food and available in many places only at a high energy cost. Climate change brings rising oceans, salt invasion and drought so millions will have nowhere to migrate.
4. Climate change	Global climate change caused by increases in atmospheric CO_2, nitric oxides and soot particulates due to combustion of the fossil fuels threatens entire species with extinction, severe weather disturbances and water and food insecurity.[118]

Consider the impacts fossil fuels cause through climate change on human health and economic vitality:

- Each year there will be new records for the hottest days and months which will increase heat-related deaths and disabilities and damage to food production.
- Increases in black soot, smog and tiny particulates will increase respiratory illnesses, dramatically reduce the quality of life and significantly diminish crop production.
- Fiercer storms caused by warmer air and ocean surface temperatures will cause more severe and frequent flooding, resulting in loss of life and property.
- Coastal recreation, businesses, fishing, tourism and homes will be inundated with rising sea levels and coastal agriculture will be ruined by sea salt invasion.

The most serious threat to human societies is the massive release of greenhouse gases from fossil fuels that accelerate global warming.

Fossil Fuels

The survival of political leaders in developing countries depends on providing affordable fuel and food for homes, farms, businesses, industry and power plants. Oil and coal are the only practical fuels available currently.

The fossil fuel of choice for powering the electrical grid is coal because coal is much cheaper than oil or natural gas and burns hotter, thereby producing more electricity. Unfortunately, burning coal adds roughly twice as much greenhouse gas pollutants as fuel oil. However, given that Australia and several other countries have enormous coal reserves and is willing to ship them to willing buyers, coal-fired power plants will be adding billions of tons of pollutants to the atmosphere.

The smog in China during the 2008 Olympics provided tangible evidence of the results from burning enormous amounts of fossil fuels. The World Health Organization estimates that diseases triggered by air pollution kill 656,000 Chinese citizens each year and polluted drinking water kills another 95,600.[119]

China already mines and burns more coal than any other country. Together, China and India control more than 20% of the Earth's immense coal reserves. Over the next 30 years, the two countries plan to open another coal-fired power plant every six days.[120] China and India are likely to add about twice as much coal-fired generating capacity as the U.S. has today. The Persian Gulf states are planning significant coal imports because even in Saudi Arabia, coal generates cheaper electricity than oil or gas.

Several countries, including China and India, are building plants to convert coal to liquid transportation fuel. Unfortunately, the conversion process also releases twice as much CO_2 and other pollutants into the atmosphere as burning fuel oils.

The only other practical alternative to coal currently is uranium, a carbon-free fuel. The world has 439 nuclear power plants operating in 31 countries. China plans to build 100 new plants in the next 20 years. By 2020, a new reactor will be starting up somewhere in the world every six days, compared with one every 17 days in the 1980s. China is building coal and nuclear plants in Pakistan, Russia, Iran, Bulgaria and India.[121] Unfortunately, global reserves of uranium are estimated to be less than 40 years.[122]

Globally, fossil fuel consumption will increase dramatically unless alternative green energy sources are found. The rising middle class in China, India and Indonesia will create significantly more demand for fossil fuels as these people demand cars which will further increase pollution. This expanding bubble of new consumers will also demand higher level foods, especially meat, that consume many times more fossil fuel, cropland and water than vegetables and food grains.

American leadership

Overconsumption of fossil fuels is self-reinforcing, not self correcting. American and global consumers want more and bigger cars and also want more fossil fuel-intensive foods. Solving the fossil fuel crisis will require both supply solutions and significant changes in consumer behavior. Inaction will bring disaster for America and the world.

Long before Americans are cold and in the dark, other impacts will be even more devastating than the unprecedented transfer of wealth to other countries to pay for oil imports. Diminishing fossil fuel supplies will increase energy costs substantially and jeopardize the America's ecosystem and economy.

Table 4.4 Fossil Fuel Impacts to the United States

Issue	Description
Food security	Escalating energy costs will devastate U.S. food production that is heavily dependent on fossil fuels for farming and crop inputs. In 2008, over 60 million Americans are already on food support. A recession or further food price increases could push 120 million Americans to food support – 40% of U.S. citizens.

Foreign debt	The $9.5 trillion U.S. foreign debt is currently being supported by countries that have the expectation the robust American economy will service the debt. America cannot afford the severe growth in the balance of payments deficit. In the past, agricultural exports partially offset oil imports.
National security	As fuel scarcity becomes more severe, the threat of disruption in production, supply, refining or distribution due to nature, politics, terrorist, weather or other factor threaten American's health and welfare, food, transportation and the US military's access to strategic fuel.
Exports	Higher fuel costs and growing food crops for fuel combined with freshwater scarcity will diminish agricultural exports significantly. In 2008, over a dozen countries ceased food exports due to food shortages. Many of the 155 countries that currently buy food, mostly food grains, from the U.S. will not be able to afford the prices American farmers and exporters will have to charge.
Water security	The distribution of water for agriculture, industry and cities consumes around 25% of America's grid energy. Higher energy costs will reduce water availability for all users – especially for farmers and the food supply chain.
Consumer prices	Consumers know that fuel price increases drive up prices for nearly all consumer goods, especially food and transportation. Fossil fuels represent the largest cost in food production. A diesel guzzling tractor gets about six miles per gallon and often travels each row in the field eight times to raise a single crop. In addition, 90% of the cost of

 synthetic nitrogen fertilizer comes from the fossil fuel used to manufacture the fertilizer.

Prior to the 1973 OAPEC (The extra "A" is for Arab) oil embargo, the price of a barrel of oil and a ton of food grain had been in sync at about and equal price for decades. Since the embargo, the OPEC alliance has orchestrated supply to double the price of a barrel of oil seven times while the price of food grains has only doubled. Rising food prices will make U.S. produced food unaffordable to both Americans and U.S. trading partners.

Fuel costs for transportation often double the cost of food. Consequently, food, fuel and transportation are closely linked. Business as usual will push fuel prices higher and create severe drag for the US economy and families. America will run out of affordable food and energy. Reducing energy cost represents the only viable way of bringing down the cost of food and transportation.

The only way to stop climate change is to significantly reduce or eliminate fossil fuel emissions of heat-trapping gases. Replacing fossil fuels with sustainable carbon-neutral fuels will make a significant difference in the health of our planet. Green solutions will be good for the U.S. economy, end dependence on foreign oil and enhance energy security.

The magnitude of fossil fuel consumption and associated pollution mean these issues need to be addressed as a world community. The U.S. may take the lead in a plan of action because America stands to gain tremendous value from:

- An end to expensive fossil fuel imports.
- Paying Americans rather than foreigners for food and fuel.
- A short and reliable food and energy supply chain.
- Ending the need to mine, transport and burn coal to power the electric grid.
- A green energy bubble that creates entrepreneurial businesses.
- Thousands of green collar jobs that cannot be outsourced.
- Sustainable and affordable energy, water and food.

Fossil Fuels

- A change from wasteful and disposable consumptive behavior to green sustainable choices that benefit the country, communities and families.

The only practical way to achieve these goals is to engineer SAFE production that does not compete with industrial foods for limited fossil resources.

Chapter 5. Fossil Food Nutrients

I cannot over-emphasize the importance of phosphorus not only to agriculture and soil conservation but also to the physical health and economic security of the people of the nation.

— President Franklin D. Roosevelt

Farmers currently use over 207 million tons of mined inorganic fertilizers annually which are sustainable only as long as all the ingredients in fertilizer are economically recoverable.[123] Fertilizer production accounts for 30% of farm energy use in the U.S. and is sustainable only as long as cheap fossil fuels are available.[124]

Fertilizers are not sustainable because they depend on fossil fuels for mining, transport and application and several fertilizer ingredients will peak in a generation. Phosphorus, copper and zinc are likely to become unavailable or unaffordable, even before the extinction of fossil fuels. Building a food supply system on industrially fixed nitrogen and mined inorganic chemicals makes our ability to feed ourselves dependent upon a non-renewable fossil fuel, limited agricultural chemicals and the wisdom, benevolence and cooperation of heads of state and multinational petroleum companies.

Fertilized soils release more than two billion tons of greenhouse gases every year, especially CO_2, methane and nitric oxide. Each acre of corn production adds 2.25 tons of CO_2 to the air plus nitric oxide that has 296 times the warming capacity of CO_2.[125]

Plants need 17 elements that are absolutely required for normal plant growth and development. Plants accept no substitutes for these elements and can take them from organic or inorganic sources. Most nutrients, especially inorganic elements, must be broken down by soil microbes before they can be absorbed by plants. Seeds may germinate and plants may grow with insufficient nutrients, but they do not prosper. In fact, plants need each of the 17 elements, Table 5.1, in order to set their fruit, which is the whole point of growing food crops.

Table 5.1 Required Nutrients for Crop Production

Macronutrients		Micronutrients	
Nitrogen	(N)	Boron	(B)
Phosphorus	(P)	Copper	(Cu)
Potassium	(K)	Chlorine	(Cl)
Carbon	(C)	Iron	(Fe)
Oxygen	(O)	Molybdenum	(Mo)
Hydrogen	(H)	Manganese	(Mn)
Sulfur	(S)	Nickel	(Ni)
Magnesium	(Mg)	Zinc	(Zn)
Calcium	(Ca)		

Nutrient need is based on the weight of the dry matter of the plant. Macronutrients requirements are 1000 mg per kilogram of dry matter or more while micronutrients are required in smaller amounts, typically less than 100 mg per kilogram of dry matter. As plants grow, they remove these nutrients from the soil. When the crops are harvested, about half the soil nutrients are lost in the plant biomass. Often less than 20% of the biomass provides usable food and the remainder (stover) is often burned or buried so the residual nutrients are lost to the field. When micronutrients are depleted and not replaced, food crops do not thrive. Crops grown in nutrient deficient soil show up to 75% less of the deficient nutrient in foods compared with fertile soil.[126] Crops lacking sufficient nitrogen or phosphorus may fail completely and crops lacking zinc may fail to set fruit.

From 1950 to 1980, U.S. grain production per acre increased about four times while the fertilizer used to produce grain increased about 30 times.[127] Since 1950, India has increased nitrogen, phosphorous and potash fertilization 219, 723 and 804 times respectively.[128] Water applications also increased at high rates because additional water improves seed germination, fertilizer absorption and plant growth.

Since 1980, U.S. grain crop yields have declined by about 1% a year because crops have limits to the amount of fertilizer and pesticides that they can tolerate. Overuse of nitrogen fertilizer can be toxic to crops. The combination of less productive fertilizer and population expansion has caused world grain production, on a per capita basis, to decline nearly every year since 1984, Figure 5.1.[129] In 2007, U.S. farmers applied about 20 million tons of fertilizers to their fields while Chinese farmers applied 48 million. Fertilizer application by Chinese farmers is expected to double by 2030.

Figure 5.1 Global grain production per person

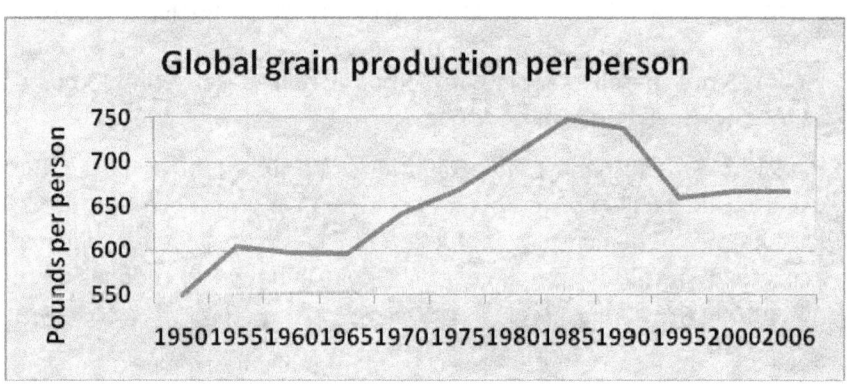

Heavy fertilizer demand predictably drives up prices. Since 1960, the price of nitrogen fertilizer has increased 540%, Figure 5.2. Both potassium and phosphate must be mined and refined and their prices have gone up 300% during the past five years. The International Fertilizer Industry Association reported that overall global consumption of fertilizer increased by an estimated 31% over the last decade, with a 56% increase in developing countries.

Figure 5.2 Fertilizer Prices – Dollars per Ton

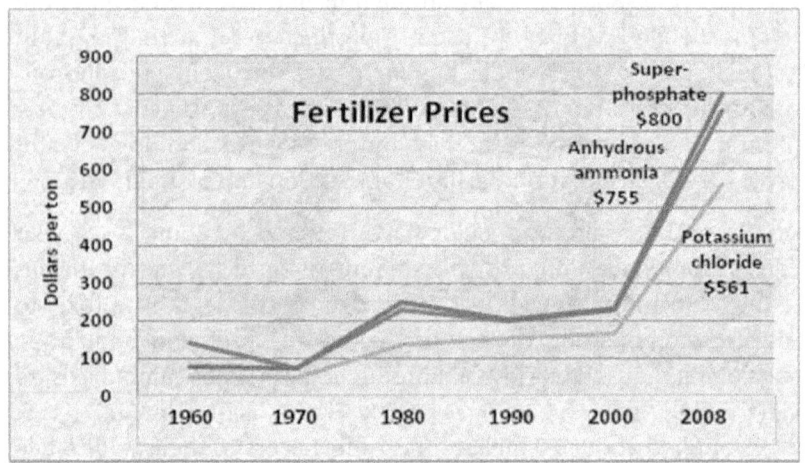

In May 2008, Senator Kent Conrad (D-N.D.), a member of the Senate Agriculture Committee, and chairman of the Budget Committee and fellow North Dakota Senator Byron Dorgan asked Agriculture Secretary Ed Schafer to investigate why prices for fertilizer products rose from $400 to $1,000 per ton in 2007. They noted that farm incomes are barely keeping up with increased farm input costs, especially escalating fertilizer prices.

Results from the Senate investigation of fertilizer prices noted the usual suspects: increases in fossil fuel prices, transportation costs, speculators, demand from biofuels, tight credit, production balance and greedy fertilizer companies. [130] Diminishing world supplies were not among the findings.

India recognized the need for government subsidies for fertilizer and began subsidies in 2005 totaling about $4 billion. Rising costs for fertilizer imports and diminishing supplies increased subsidy costs to $22 billion in 2008, which has prompted calls to reform the program that India depends on to maintain its food supply.[131] India has also reported severe problems with agricultural chemical pollution because monsoon rains move tremendous amounts of soil and embedded chemicals. A high number of deaths and disabilities among farmers have also been reported in India and other countries because many farmers cannot read the instruction labels on fertilizers and poisonous agricultural chemicals.

Manure

For centuries farmers practiced organic farming to improve crop yields with manure and compost. However, manure is a relatively inefficient fertilizer and creates several problems. Manure contains only a modest amount of nitrogen so heavy application is required to replace the nitrogen removed by the prior crop. On modern farms, manure animals are often raised thousands of miles from field crops – which makes transporting heavy manure impractical. Even if meat and dairy animals were close to fields, there are far too few animals to supply sufficient manure for vast croplands.

Plants growing in the wild are fertilized by natural organics (last year's plant remains plus organics left by birds and animals). Farmers found they could increase yields of food crops by providing additional fertilizer, typically manure, cinders (potash) and iron slag, to improve crop yields to promote plant growth and development.

Organic farmers apply manure rather than synthetic or mined fertilizers. Manure contains about 80% of the original plant nutrients but most of the nitrogen evaporates from manure through ammonia volatization.[132] Manure must be plowed into the soil to retain the nitrogen, which takes considerable energy, disrupts soil ecology and promotes erosion. Evaporating ammonia from manure creates nitric oxides, which contribute to global warming and depletes ozone protection in the upper atmosphere.

Drugs create another problem with manure as fertilizer. For 60 years, meat producers have fed antibiotics to farm animals to increase their growth and prevent infections. About 70% of all antibiotics produced in the U.S. (nearly 25 million pounds a year) are fed to cattle, pigs and poultry according to the Union of Concerned Scientists.[133] Feeding pharmaceuticals to animals sustains a growing demand for meat, but it generates public health fears associated with the expanding presence of antibiotics in the food chain. About 90% of the drugs end up being excreted either as urine or manure. Food crops such as corn, potatoes and lettuce absorb and concentrate antibiotics when grown in soil fertilized with livestock manure.[134]

Most land plants can absorb nitrogen only through their roots. Even though the atmosphere is nitrogen rich, plants need added nitrogen in the soil in a form they can absorb. To solve this problem, organic farmers grow green manure crops such as legumes (clover, vetch, alfalfa or beans) in rotation with crops such as food grains. Green manure crops fix atmospheric nitrogen in their roots in a form that plants can use. Cover crops also add other nutrients, organic matter to enrich soils, suppress weeds and moderate erosion. Organic farmers also may compost various agricultural materials to provide natural fertilizer. While organic farmers may use less fossil energy than other farmers, they are still heavily dependent on fossil fuels.

Fertilizer use

Corn farmers in the U.S. apply about the following per acre:
- Nitrogen (N): 130 pounds
- Phosphorous (P): 60 pounds
- Potassium (K): 65 pounds
- Lime 450 pounds (Calcium neutralizes soil acid and improves soil bacteria activity.)[135]

Corn plants, like many food crops, are fussy about how they take their nutrients and cannot absorb fertilizers until soil microbes have broken down the elements. Consequently, many farmers use specially prepared liquid fertilizers and "side dress" the plants by spraying or injecting the liquid into the roots. Naturally, this creates more cost and energy consumption than simply spreading manure or another high nitrogen fertilizer. Fertilizers are inefficiently absorbed by corn, even in liquid form, so farmers often apply twice as much as the plant actually needs. Farmers who plant corn in the same field in consecutive years may have to more than double the fertilizer applied due to nutrient loss.

Fossil Food Nutrients

Organic fertilizers are composed of naturally occurring organic matter and inorganic fertilizers are made of inorganic chemicals or minerals. When fossil fuels were cheap and mineral deposits such as peat and potash were plentiful, fertilizers were cheap. Much of the productivity associated with the Green Revolution that brought substantial improved crop productivity came from adding more critical fossil resources; especially water and fertilizer. Two graphics illustrate the fertilizer story, fertilizer use and fertilizer use by crop. Fertilizer use Figure 4.2 since 1964 indicates added nitrogen has increased 660%.

Figure 5.2. Fertilizer use in the U.S. in nutrient tons

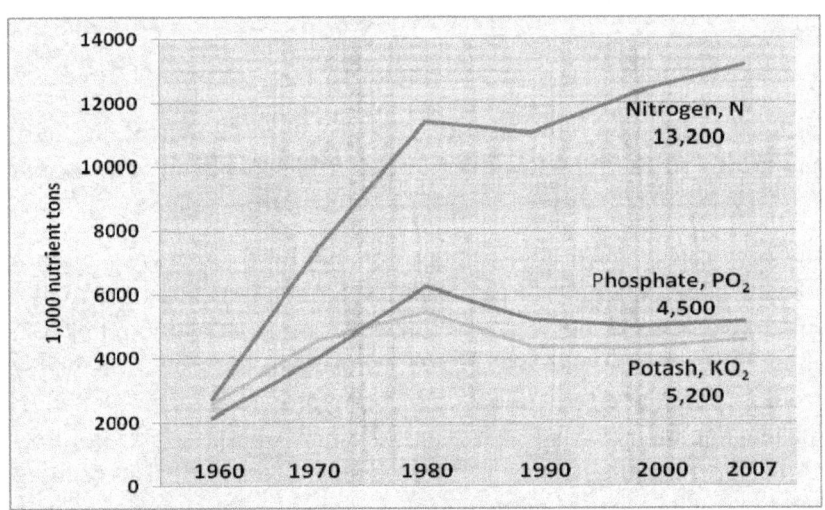

Nitrogen use by various crops, Figure 4.3, shows that corn production was the primary recipient of added nitrogen. Corn consumes substantially more fertilizer than other subsidized crops which makes corn the major contributor to water pollution. Corn uses more than 10 times more nitrogen per acre than soybean production.

Figure 5.3. Fertilizer Use by Crop

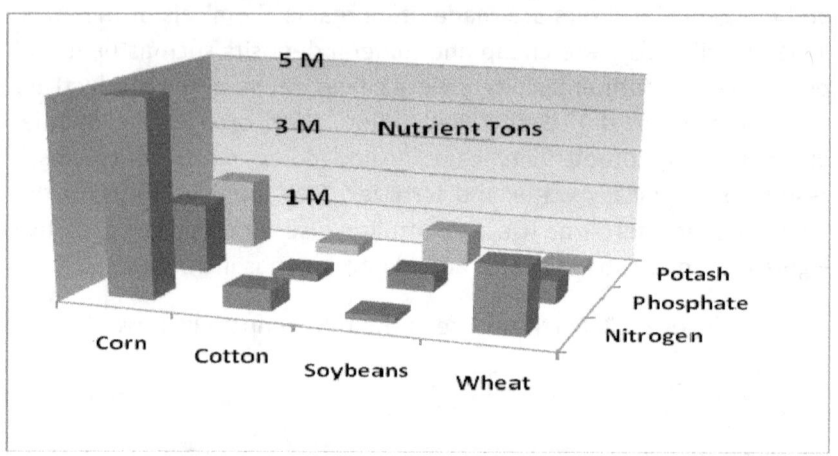

Each U.S. corn crop removes more than 6 billion pounds of nitrogen, hay grown to feed livestock removes 7.4 billion pounds and wheat extracts another 2.4 billion pounds of nitrogen from soils.

The corn rush caused by escalating ethanol production causes more corn to be grown on or near environmentally sensitive land that is vulnerable to sediment and chemical invasion from erosion. In 2007, about 20% of U.S. corn was grown on acreage designated as highly erodible land and additional corn was produced near wetlands, shallow aquifers, rivers, and streams.[136] Since American farmers apply 13 million nutrient tons of nitrogen, five million tons of phosphate and four million tons of potash fertilizer annually, it is easy to see how fertilizers relentlessly invade water sources.[137]

China subsidizes fertilizer manufacturing. The average grain yield per acre doubled between 1977 and 2005, while nitrogen fertilizer use increased 271%. Nutrients applied to typical fields in China far exceed those in the U.S and northern Europe – and much of the excess fertilizer is lost to the environment – degrading soil, air and water.[138]

Nitrogen

Nitrogen (N) enables living cells to live and prosper. Nitrogen provides a building block of amino acids and nucleic acids, essential component of both RNA and DNA, the genetic blueprints that pass genetic characteristics from one generation to the next. Nitrogen serves several key roles in metabolic processes involved in the synthesis and transfer of energy and enables the use of chlorophyll, the green pigment in plants that generates energy and growth. Nitrogen speeds plant growth, increases seed and fruit production, and improves the quality of leaf and forage crops.

Nitrogen plays a key role in plant, animal and human nutrition and biological molecules including proteins which are made from amino acids and nucleic acids. Plants use nitrogen in chlorophyll molecules, which enables photosynthesis, essential plant growth and vitality.[139] Therefore, nitrogen is critical for all living things. Nitrogen typically makes up around 4% of the dry weight of plant matter and roughly 3% of the human body.

Synthetic nitrogen is a fossil resource not because of the nitrogen, which is plentiful since it makes up 78% of the atmosphere, but due to the fact that sources of hydrogen are limited and energy costly. Hydrogen is necessary to transform N into a form that can be absorbed by plants. Triple chemical bonds between the two atoms of the gaseous N_2 molecule render atmospheric nitrogen inert or incapable of linking with atoms of other elements. Nitrogen gas abstains from participating in chemical reactions. Before plants can make use of atmospheric nitrogen, it must be "fixed" (the process of breaking N_2 atoms apart) so that the nitrogen atoms can combine with elements that may be bioavailable to plants.

Only two natural phenomena can fix nitrogen: lightning and bacteria. Lightning's powerful flashes cleave the nitrogen molecule bonds. The free nitrogen atoms combine with oxygen and fall to Earth as nitrates (NO_3) – a form of nitrogen that plants can absorb. Lightning contributes around 10 million metric tons of nitrogen per year for use by plants and animals.[140]

Special bacteria generate energy by oxidizing or "burning" carbohydrates to fix atmospheric nitrogen. Rhizobium forms symbiotic relationships with members of the legume plant family such as clover, alfalfa and beans. The bacteria convert molecules of nitrogen trapped in air pockets in the soil to ammonia (NH_3), which other types of bacteria then convert into nitrates. Microorganisms add around 140 million metric tons of fixed nitrogen to the soil every year.

Once fixed nitrogen has been taken up by plant roots, the plant uses it to create biomass that passes up the food chain to animals where it enables growth. When excreted, the nitrogen returns to the soil where it acts as a fertilizer with dead plant and animal material. Bacteria decompose the fallen organic materials and recycle the nitrogen to the nitrate form which enables reuse by plants.

Until the 20th century, farmers seeking to provide sufficient nitrogen for their crops turned cover crops, manures or compost, into their soils and relied upon soil microbes to make the nitrogen in these materials available to plants' roots.

Fritz Haber changed agriculture in 1909 when he developed a high-temperature, high-pressure (high energy) process in his laboratory to fix atmospheric nitrogen. Carl Bosch expanded Haber's process to a factory scale. Known as the Haber-Bosch process, industrial fixation of nitrogen combines atmospheric nitrogen and hydrogen into ammonia – the basis for synthetic to make nitrogen fertilizers, NO_x. (The x denotes that various N molecules use various numbers of O atoms.)

The Haber-Bosch process creates synthetic nitrogen fertilizer, anhydrous ammonia, but consumes 33,500 cubic feet of natural gas (methane) to produce one ton of anhydrous ammonia (82-0-0). The zeros are placeholders for phosphorus and potassium. Anhydrous ammonia is then used to manufacture dry nitrogen fertilizers, such as ammonium nitrate (34-0-0) and urea (46-0-0). Alternative hydrogen sources such as naptha, a hydrogen rich hydrocarbon refined from petroleum, are also fossil resources.

Animal waste contains copious amounts of nitrogen, usually in the form of urea, uric acid and ammonium compounds. Nitrates are extreme polluters because their high solubility enables rain and irrigation to quickly rinse nitrogen out of soils. Heavy nitrogen flows pollute streams, lakes, groundwater and oceans. Oceans now have more than 400 dead zones caused by nitrogen and phosphorus pollution where dissolved oxygen has been depleted and all aquatic life has died. A recent report in *Science* found that dead zones were increasing in size about 10% a decade.

Elevated nitrates in groundwater create a set of health risks including methemoglobinemia, blue-baby syndrome, where nitrates interfere with blood-oxygen levels in infants. A 2007 Iowa Department of Natural Resources report indicates 274 Iowa waterways were seriously polluted.[1] Fertilizer run-off from corn causes such a problem that Iowa plans to build the largest and most expensive ($455 million) nitrate removal plant in the world.[2]

Wetlands typically nurture slow growing, low- nitrogen plants such as ferns that have low nitrogen needs because nitrogen is sparse in their ecosystems. When runoff water high in nitrogen invades wetlands, fast growing opportunistic overwhelm existing plants like ferns. Nitrate-enriched water leads to entrophication, the growth of cyanobacteria, which results in dead zones depletion of dissolved oxygen and causes death to fish and all other water creatures.

Synthetic nitrogen

Synthetic nitrogen fertilizers are made using the Haber-Bosch ammonia synthesis that converts N_2 from the air into a plant usable form, anhydrous ammonia.

$$\text{Air } (N_2) + \text{Natural Gas } (CH_4) = \text{Anhydrous ammonia } (NH_3)$$

Air and natural gas are combined at high temperature, about 900° F, and pressure, between 400 and 1,000 atmospheres, to create anhydrous ammonia gas. Natural gas follows a two-step process where it reacts with the atmosphere to supply hydrogen to the reaction and then to create the high temperature and pressure necessary for the process to take place. For decades, the U.S. was the largest exporter of fertilizers but now, largely due to ethanol

---

production, farmers must import more than half the U.S. consumption.[3]

Manufacturing one ton of anhydrous ammonia fertilizer requires huge amounts of natural gas, creating about 90% of the fertilizer cost.[4] When natural gas prices were $2.50 per thousand cubic feet, the natural gas used for one ton of fertilizer cost $84. When natural gas prices rose to $7 per thousand cubic feet, the cost of natural gas alone cost $235, an increase of $151. Transportation cost has recently been added to the cost of nitrogen fertilizer because ocean shipping fees increased 300% in 2008.[5]

Synthetic nitrogen fertilizer, compared with organic, increases crop yields per acre and the physical size of vegetables. However, these larger vegetables typically have significantly reduced ascorbic acid (vitamin C) content and diminished other vitamins and minerals. Synthetic nitrogen may also increase nitrate concentrations to dangerous levels in edible plants.

Phosphorus

Phosphorus is an irreplaceable ingredient of life but probably the most critical limited fossil resource for agriculture, after water. Phosphorus, from Greek meaning light-bearer, is critical for all living cells. Phospholipids form and sustain cell membranes. Phosphorus serves as the key structural component of DNA and RNA. Phosphorus provides shape for DNA, which provides the blueprint of genetic information contained in every living cell. A sugar-phosphate backbone forms the helical structure of every DNA molecule. The element also regulates ATP (Adenosine-5'-triphosphate), which is the main energy storage and transfer molecule in cells. It is also necessary for the formation and maintenance of bones and teeth in animals and humans. The human body contains about 650 grams of phosphorus, most of it in our bones.[6] About 20% of the human skeleton and teeth are made of calcium phosphate, Ca(H2PO$_4$)$_2$.

The phosphate ion combines with various atoms and molecules within living organisms to form many different compounds essential to life. Unfortunately, phosphorus is one of the most chemically reactive nutrients and readily transforms to a variety of phosphorus compounds with no bioavailability to plants.

Phosphorus is highly reactive with air and other oxygen-containing substances and is not found free in its elemental form in nature. Phosphorus exists in small concentrations in many different minerals, which make it expensive and energy intensive to mine. It occurs as a charged group of atoms or an ion made up of a phosphorus atom and four oxygen atoms (PO_4) and carries three negative charges. Miners do not mine phosphorus; they mine phosphate minerals and then use additional energy to extract the phosphate.

Phosphorus deficit

Phosphorus has been accurately described by the superb scientist and science-fiction writer, Isaac Asimov, who called it "life's bottleneck." He observed that if you have a miniscule amount of something but only need a tiny, tiny bit of it, then that is less critical than if you have a fair amount of something but you need a lot of it. Asimov noted that some mineral elements are more common in organism bodies than in the surrounding environment. Therefore, the organism needs to concentrate that element and the degree of concentration of that element in the organism's body provides an indication of two things:

1. How much organisms need that element.
2. How available the element is in the environment.

Asimov calculated that phosphorus composes about 0.12% of typical soil yet gets concentrated in an alfalfa plant's body at about 0.7%.

Therefore, the concentration factor for phosphorus is about 5.8. The next highest concentration factor is sulfur at 2 and chlorine at 1.5. Therefore, if there are more and more organisms needing mineral elements, or if the living ecosystem is more and more depleted of its resources, the mineral element that will come into short supply first is predictable: phosphorus. Unfortunately, Asimov was right and phosphorus will probably be the first critical fossil resource to run out which will devastate the global food supply.

The natural biochemical cycle recycles phosphorus back to the soil 'in situ,' from dead plant matter. Land ecosystems use and reuse phosphorus in local cycles about 46 times. The mineral, through weathering and runoff, makes its way to the ocean, where marine organisms may recycle it some 800 times before it passes into sediments. Over tens of millions of years, tectonic uplift may return it to dry land.[7]

The Chinese have used human excreta ('night soil') as a fertilizer for thousands of years. The Japanese have used human excreta since the 12th century.[141] In Europe, soil degradation and recurring famines during the 17th and 18th centuries created the need to supplement animal and human excreta with other sources of phosphorus. In the early 19th century, England imported large quantities of bones from other European countries for use a fertilizer[142]

Throughout 99% of human agricultural history, crop residues and organic matter (animal manure and human wastes) were returned to the field. Around the mid 19th century, the use of local organic matter was replaced by phosphorus material from distant sources. The mining of guano, bird droppings deposited more than previous millennia as well as phosphate-rich rock began.[143] Guano was discovered on islands off the Peruvian coast and later on islands in the South Pacific. World trade in guano grew rapidly but it relied on a limited resource which declined by the end of the 19th century.

Fossil Food Nutrients

Phosphate rock was seen as an unlimited source of concentrated phosphorus and the market for mineral fertilizers developed rapidly.

Flush toilets became popular in the 1840s and contributed significantly to the phosphorus deficit. Toilets flushed human waste into municipal waste streams and the phosphorus is lost to the fields. These actions spurred protests among intellectuals that farmers were being robbed of human manure. Among the protestors was Victor Hugo, who wrote in *Les Miserables*:

> Science, after having long groped about, now knows that the most fecundating and the most efficacious of fertilizers is human manure. The Chinese, let us confess it to our shame, knew it before us. Not a Chinese peasant goes to town without bringing back with him, at the two extremities of his bamboo pole, two full buckets of what we designate as filth. Thanks to human dung, the earth in China is still as young as in the days of Abraham. Chinese wheat yields a hundredfold of the seed. There is no guano comparable in fertility with the detritus of a capital. A great city is the most mighty of dung-makers. Certain success would attend the experiment of employing the city to manure the plain. If our gold is manure, our manure, on the other hand, is gold.[144]

As the combination of population and trade in food products increased, insufficient amounts of nutrients were returned to the soil. By the end of the 19th century, processed mineral phosphorus fertilizer was routinely used in Europe and its use grew substantially in the 20th century.

The strategic importance of phosphorus was not lost in America. In a message to Congress in 1938, President Franklin D. Roosevelt underscored the importance of phosphate to agriculture and people.

> The phosphorus content of our land, following generations of cultivation, has greatly diminished. It needs replenishing. I cannot over-emphasize the importance of phosphorus not only to agriculture and soil conservation but also to the physical health and economic security of the people of the nation. Many of our soil deposits are deficient in phosphorus, thus causing low yield and poor quality of crops and pastures.[145]

Today, the USGS estimates global phosphate mining extracts about 22 million metric tons a year. There is no artificial substitute for phosphorus in agriculture. As world reserves of this critical natural resource diminish, prices will increase. Scientists at Linköping University in Sweden predict peak phosphorus will occur in 20 years and poses a serious threat to agriculture as global reserves of high-quality phosphate rock go into terminal decline.[146]

World governments are beginning to recognize the strategic value of phosphorus, which is also used heavily in the munitions and chemical industries. India is running low on matches and fireworks as factories run short of phosphorus. The Brazilian government is debating whether to nationalize privately held mines that supply the fertilizer industry.

China, the largest producer and the largest user, recently imposed a net zero export policy on phosphorus. Beijing levied a 135% tariff on phosphate rock exports. Naturally, Europe and India are upset because they are reliant on phosphorus imports. U.S. phosphorous production has dropped 20% over the last three years, forcing more phosphorus imports from Morocco.

Phosphorus availability

Modern agriculture depends completely on the regular application of phosphate fertilizer derived from mined rock. Modern agriculture harvests crops prior to their decay phase (which would replace extracted phosphorus) and transports food all over the world. About half the phosphorus in the soil is lost with the harvest as food goes from the field to fork and the phosphorus winds up in human or animal waste streams.[147] Another third or more is lost to erosion from wind and water.

Phosphorus availability represents a serious world crisis because plants require phosphorus to grow and the lives of animals and humans depend on food crops.

Table 5.2. The Global Phosphorus Crisis

Issue	Description
Supply	Phosphorus is a limited-supply element on the periodic table (atomic number 15) that cannot be manufactured and neither plants nor animals accept any synthetic substitutes.
P blindness	Phosphorus blindness occurs because leaders and policy makers are blind to peak phosphorus while they debate peak oil and global climate chaos.
Critical for food production	About 90% of the global demand for phosphorus goes to food production, currently around 148 million tons of phosphate rock per year.[148]
Depend on cheap fossil fuels	Phosphate mining, refining, packaging, transportation, storage and application to fields depend on cheap fossil fuels because the supply chain is long, expensive and energy intensive.
Global demand	Global demand for phosphorus is forecast to increase by around by 4% annually with about two-thirds of this demand coming from Asia, where both absolute and per capita demand for phosphate fertilizers is increasing.[149]
Peak phosphorus by 2030	Phosphorus is a non-renewable resource and current estimates estimate economically recoverable reserves will be depleted in 50 years with peak phosphorus occurring by 2030.[150]

Quality declining	While the timing of the production peak may be uncertain, the fertilizer industry recognizes that the quality of existing phosphate rock is declining and cheap fertilizers will soon become a thing of the past. The average grade of phosphate rock has declined from 15% P in 1970s to less than 13% P in 1996.[151]
Increasing extraction cost	After peak phosphorus, the remaining potential reserves will be more costly to extract and more will have to be mined due to the lower quality rock.
Price spikes	Increased demand from food production and biofuels in 2007 pushed the price of phosphate rock up by more than 700% to $367 per ton.
U.S. in short supply	U.S. phosphorous production has dropped 20% over the last three years forcing phosphorus imports from Morocco. The U.S., historically the world's largest producer, consumer and exporter of phosphate rock, has only 25 years left of domestic reserves and now imports significant amounts of phosphate rock from Morocco.[152] The U.S. imports considerable phosphate rock because much of the U.S. phosphate rock is not as readily absorbed by plants as imported rock.
Global demand	Global demand for phosphorus is forecast to increase by around by 4% annually with about two-thirds of this demand coming from Asia, where both absolute and per capita demand for phosphate fertilizers is increasing.[153]
Africa's role	Ironically, the African continent is simultaneously the world's largest exporter of phosphate rock and

the continent with the largest food shortage. If Morocco, which controls 40% of global supply followed China's net zero export strategy, world food production would face chaos.

Only a few mines

Phosphate rock reserves are controlled only a few of countries – Morocco, China, South Africa and the U.S. – while Western Europe and India are totally dependent on imports.

The U.N. has condemned international trade that depletes North Africa's phosphorus reserves and some countries have instituted policy to reduce North African imports.[154] Several European countries have begun recycling and phosphorus management programs. Half the phosphorus excreted by humans flows out as urine. Swedish scientists have developed toilets that can recover phosphorus from human wastes. Two municipalities have mandated that all new toilets must be urine diverting.[155]

Urine diverting toilets have two holes that separate urine from solids so the liquid waste can be recovered and recycled on gardens and fields. Compost sources commonly recommend gardeners save their urine and pour it on their compost piles because it had so much phosphorus.

The other half of human consumed phosphorus lies in solid waste. It is energy expensive to recycle and difficult because residual biosolids are contaminated with many pollutants, especially heavy metals such as lead and cadmium, which leach from old pipes. When contaminated fertilizers are applied to fields, heavy metals enter the food crops.

Phosphate demand

Several factors have increased demand for phosphorus which drove up the price of phosphate rock in 2007 by more than 700% to $367 per ton:

- Biofuels, especially corn, added to the phosphorus consumption.
- Increasing populations consume additional phosphorus in the foods they eat.
- Increasing "new consumers" who move from a vegetarian diet to meat and dairy multiply the consumption of phosphorus used to produce meat. Beef, mutton or pork, for example, uses roughly 8 times more phosphorus per pound than vegetables or grains.

Phosphate rocks may look similar but they may deliver different nutrient values because the phosphate is combined with different elements that may have only partially bioavailability (absorbable and useful) to plants. The African product provides far more available phosphorus than American rock. Finely ground American phosphate rock, in acid soils, becomes slowly available after the second year and in alkaline soils, is practically worthless.[156]

Open-pit phosphate mine in Idaho

The U.S. has phosphate mines in Idaho, North Carolina and Florida. Several environmental groups are trying to close Idaho's mines due to the considerable risks to the health of the environment, communities near the mining and processing sites from selenium contamination. Phosphate mining operations dump excess soil, rock and other material on the mine sites and leave it exposed to wind, rain and snowmelt. Selenium and other hazardous compounds leach into groundwater and streams where aquatic life absorbs the selenium. Plants concentrate selenium and are consumed by aquatic insects and fish, which are eaten by other animals, birds and people. Toxins increase each step up the food chain, bioaccumulation, and the risk of poisoning increases.

Potassium

Potassium, like nitrogen and phosphorus, is essential to all forms of life – human, animal and plant. The word potassium came from the word potash because potassium was first isolated from potash. Potash was the char left in the pot when farmers burned waste materials in a pot and spread them on their fields to improve crop productivity.

Potassium provides a critical function to all types of cells by providing transport functions and volume control within the cell. Inside each cell there are a large number of proteins and other organic compounds that pass in and out of the cell in the sodium-potassium pump. Potassium is essential for critical plant functions such as photosynthesis, protein formation, water storage and nutrient transport.

Potash, the principal source of potassium fertilizer, is not in short supply but is expensive and energy intensive to mine, package and transport. As spike in fossil fuels cost or availability would severely restrict potash availability. Potash is mined in Saskatchewan, California, New Mexico, Germany and the Dead Sea. The Saskatchewan Mining Company mines potash three thousand feet below the surface through frozen rock.

PotashCorp offers a description of potash mining which is more energy intensive than mining coal.[157]

Miners begin their day by travelling 3,281 feet straight down into the earth – two thirds of a mile. The mine has two vertical shafts that are about 20 feet across; one for moving people and one for moving ore. Miners reach the bottom and then travel about four miles by jeep through underground tunnels to their worksite. They mine the ore by machine and then it is transported through a maze of tunneled roads and brought to the surface to be milled. Potash miners use four-rotor continuous boring machines. Miners can recover up to 882 tons of rock per hour making paths 26 feet wide and 28 feet high. Potash reserves are sufficient for decades but deeper mines consume prodigious amounts of fossil energy.

Other fossil nutrients required for food production are also in short supply and countries are buying stakes in mines to assure access to key nutrients. China's third largest zinc producer, Zhongjin, bought a 50.1% stake in Australian zinc miner Perilya in January 2009.[158] Wealthy speculators from several countries, including Saudi Arabia, are buying mines that produce nutrients needed in food production.

Pesticides

Even though farmers apply more than 2.5 million tons of pesticides worldwide, more than 40% of all potential food production is lost to insect, weed, and plant pathogen pests prior to harvest. After harvest, an additional 20% of food is lost to another group of pests.[159] The FAO reports that pesticide application results in an estimated 26 million human poisonings annually with 220,000 fatalities globally. Pesticides are energy intensive to make and cause an additional $10 billion a year in environmental and societal damage in the U.S.[160] Pesticides adversely impact human, animal and livestock health, elevate pest resistance, cause crop pollination problems which reduces crop productivity and damage or destroy honeybees and butterflies as well as birds, fish and other wildlife. Herbicides and fungicides share the same problems.

Pesticides used in corn production tend to be more environmentally harmful than for other plants because corn attracts an exceptionally broad spectrum of pests due to the high sugar in the kernels. Unfortunately, less than 1% of the pesticide applied reaches the target pest.[161] The FDA reported that at least 53 carcinogenic pesticides are applied in massive amounts to food crops. The EPA considers 90% of all fungicides, 60% of all herbicides, and 30% of all insecticides carcinogenic. These chemicals dissolve in water and migrate with run-off into waterways and groundwater. Herbicides and pesticides have fatal effects for creatures in wetlands and result in four million acute human poisonings a year in the U.S.[162] Exposure to pesticides in drinking water has been linked to a wide range of adverse health effects in humans and animals including cancer, Parkinson's disease, immune system depression, and endocrine disruption. Farms produce the large majority of water pollution.

Most farmers consider herbicides necessary to control competing weeds, especially since the genetically engineered corn plant has trouble competing with opportunistic weeds. Consequently, U.S. farmers apply more than one million tons of herbicides each year.[163] Farms apply more than 77 million pounds annually of the herbicide Atrazine, a known endocrine disruptor.[164] Atrazine does its job and kills weeds but does significant collateral damage. Low doses of this endocrine disruptor cause developmental harm by interfering with hormonal triggers in the development of an organism. Atrazine results in sexual abnormalities in fish, amphibians and reptile populations including hermaphroditism (having both male and female organs).[165]

SAFE production

Solving the fossil fuels challenges requires an energy source that can be produced with the following constraints.

Table 5.3 Design Constraints for a Supplemental Food Supply

Factor	Constraint
Fossil independent	May be produced with no or minimal fossil resources.
Sustainable	Continuous production for generations.

Carbon neutral	Adds no new CO_2 to the atmosphere.
Clean energy	Adds no new greenhouse gases to the atmosphere or pollution to soils or water.
Secure	Short, reliable supply chain and low transportation cost.
Minimal waste	Wastes minimal energy in producing food and energy. Food grains produce more plant waste than food or energy.
Robust	Produces in variety of growing conditions. Grows effectively in spite of global warming and climate chaos.
Efficient	Produces more energy than it consumes.
Ecologically positive	Creates no or minimal pollution of soils and water. Uses no or minimal herbicides or pesticides.
Resource efficient	The plant requires minimal inputs and consumes inputs that are plentiful and cheap.
Efficient human inputs	Minimizes human labor, human energy, risks and time.
Really, really easy to grow	Practically idiot proof because millions of farmers will want to learn how to grow this food and energy source.

The fossil food challenges, especially supply, affordability and pollution are likely to make fossil food production extinct within 50 years. Available and affordable freshwater and fossil nutrients may make fossil agriculture impractical before fossil fuels end.

Fossil Food Nutrients

The other challenges are external to agriculture but pose equal significant threats to the modern food supply. Population expansion, global climate chaos, political and social issues and monocropping each put severe strain on industrial agriculture.

Chapter 6. External Factors

Humanity is approaching a crisis point with respect to the interlocking issues of population, food, natural resources and sustainability.[166]

– The academies of sciences from 58 countries.

Many birds and animals are energy frugal in the sense that they slow activities when food becomes scarce. Coyotes build a den but skip whelping in lean years when food is sparse. Why waste energy on propagation when energy is needed to find food? Modern humans are not as smart as coyotes and continue to expand population beyond the natural carrying capacity of food supplies.

Low and stable global food prices, due to increases in agricultural productivity leveraged on fossil inputs, have created an expectation that agriculture can continue to provide sufficient food for expanding populations. While population growth has expanded exponentially, fossil resources have diminished. Several consensus reports have estimated the carrying capacity of the Earth and it seems to be around the world population in 1980, 4.5 million.[167] The *Population Summit* report documents the serious status of vital resources needed to support human life.[168] Science has not computed the carrying capacity of Earth but the report makes clear that current practices are not sustainable. The report recommends local solutions give way to more cooperative global solutions to support food security and human life. Abundant food qualifies as a needed global solution.

Much of the world population growth to 6.8 billion has occurred in since 1950, enabled by to a cheap food supply provided by Industrial agriculture dependent on fossil cheap fuels. The Black Agricultural Revolution based on fossil fuels transformed agriculture as world grain production tripled but world population increased by four billion, Figure 5.1.

Figure 6.1 Population Expansion

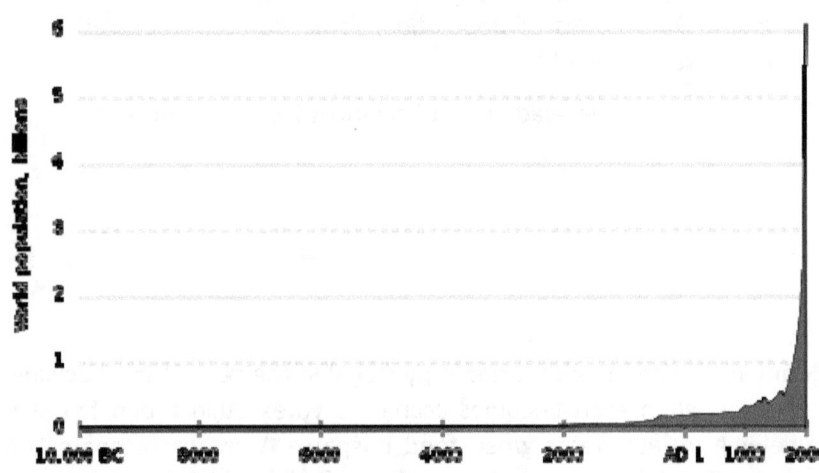

To keep up with human population growth expected to be 9 billion, more food will have to be produced worldwide over the next 50 years than has been during the past 10,000 years combined. The Earth simply has insufficient fossil resources to support substantially higher food production using fossil agriculture.

Even though food productivity increased with the fossil food production, the number of hungry people also increased because staying competitive in fossil farming requires buying expensive inputs and wealthier farmers usually win over the poor. In farming, as in buying food, the poor pay more but get less. Poor farmers cannot afford to buy fuel, seeds, fertilizer, agricultural chemicals and other inputs in volume so they forego discounts. They cannot hold out for the best price for their crops so they must take less for their production. The poor often cannot afford to pay for irrigation and they typically do not have access to credit or microloans. The poor generally also lose to wealthier producers for government subsidies and other benefits.

Poor farmers who are unable to farm due to input prices are unlikely to find adequate employment to compensate for the loss of their farming livelihoods, causing their families to go hungry. The root cause of hunger is not simply a shortage of food but the ability to afford food or access to food production inputs. Unfortunately, with peak of oil and other agricultural chemicals, those inputs will get even more expensive and further out of reach for those with the least means. Fossil agriculture creates severe social inequity.

Hungry rural populations pushed out of farming have migrated to cities where they became dependent on others for food. In 1800 only 3% of the world's population lived in cities but over half live in cities today. In 1950, only 83 cities had populations exceeding one million but today 468 cities have more than a million.[169] The U.N. forecasts that the current urban population of 3.2 billion will rise to nearly 5 billion by 2030, when three out of five people will live in cities.[170] Rural to urban migration not only results in fewer food producers but more energy consumption because urban dwellers consume about four times as much energy as their rural relatives.[171]

Climate change

Many scientists, including James Hansen at NASA, believe that global climate change is accelerating and may be approaching a tipping point where climate change acquires a momentum that makes it irreversible. The consensus is that we may have only a decade to turn the situation around before this threshold is crossed.[172]

Global warming that included drought, hot, dry winds, wildfires and fierce storms was largely responsible for food price spikes in 2008 that caused food riots in 33 countries. These insurrections disrupted national economies, spurred food theft and resulted in hundreds of deaths. Several countries created policies that prohibited food hording, waste and even food exports.

In anticipation of more food shortages due to global warming and population increases, a parade of over 50 credible voices has called for a doubling of the world's food supply in the next 30 years, including:

- Robert Zoellick, World Bank president[173]
- Ban Ki-moon, United Nations Secretary General[174]
- John Beddington, the United Kingdom's Chief Scientific Adviser[175]
- LaMar Lemmons, Michigan State House of Representatives[176]
- Hugh Grant, Chairman, President and CEO of Monsanto[177]

These world leaders may be unaware that agriculture may have already peaked.[178] World food production may not have the capacity to expand at all or even to sustain current production with conventional agriculture.[179] Economists and even ag-economists fail to factor into their business models the cost of agriculture's self-destruction, over-consumption of fossil resources and climate shocks. Nicholas Stern, former chief economist at the World Bank, examined the cost of failing to incorporate the climate change costs and burning fossil fuels and concluded the cost would be in the trillions of dollars.[180] The convergence of many factors may make fossil intensive food production unsustainable, even with expected improvements in agriculture technologies.

Modern crops are not built for climate change, especially heat, drought and salt. Food crops were hybridized over the past 11,000 thousand years to grow within a relatively narrow temperature range, absent shocks from climate change. Average temperatures in food growing regions have increased $1°$ F since 1970 and 24 of the warmest years on record have occurred since 1980.[181]

Unusual summer weather in the U.S. in 2009, rain, heat and wind, combined to cause severe tomato blight throughout the Northeast and Mid-Atlantic States that threatened the entire tomato crop.[182] The "late blight," that could jump to potatoes, caused white, powdery spores, large olive green or brown spots on leaves and fruit and open lesions on the stems. Each lesion produced hundreds of thousands of infectious spores that spread on the wind, creating an explosive infection to other fields. Fungicides may protect unaffected plants from disease, but there is no cure for late blight. Many farmers had to plough their fields under, losing their entire crop. The late blight is similar to the potato blight that caused the Irish famine in 1845 that caused starvation of over one million people.

Too much heat devastates food grain yields. As temperature rises, the rate of photosynthesis increases to about 68° F and then plateaus up to 95 ° F. Photosynthetic activity declines above 95° F and stops at 104° F.[183] Rice, wheat and corn cannot pollinate above 104° F, which leads to crop failure. Combinations of heat, dry winds and insufficient soil moisture create partial pollination problems well below 104° F. Since pollination occurs only after the plant is fully grown, farmers must invest all their resources for an entire growing season before a heat spike ruins their crop.

Heat stress causes plants to curl their leaves to get less solar exposure and the stomata to close on the underside of the leaves in order to decrease moisture loss. Both actions disrupt photosynthesis and send the plant into thermal shock. David Lobel and Gregory Asner analyzed 16 years of corn and soybean data across 618 countries and concluded that each 1.8° F rise in temperature, over normal, caused a yield decline of 17%.[184] Some crops, such as rice, cannot pollinate if night temperatures stay above 80° F. Heat stress causes yield drag for most crops of 15—25%, but severe heat stress from only a few extra degrees may destroy an entire crop. Heat stress breaks crop vitality and leaves the plants more vulnerable to disease, pests and fungus.

The heat wave in Europe in 2003 was only 2° F (3.6°C) above the long-term climatology averages but led to the deaths of an estimated 52,000 people.[185] Italy experienced a record drop in maize yields of 36% from a year earlier, whereas in France maize and fodder production fell by 30%, fruit harvests declined by 25%, and wheat harvests declined by 21%.[186] Temperature spikes, both high and low, can be extremely destructive especially at critical phases such as germination, early growth, pollination and fruiting. High winds associated with fierce storms can destroy a crop or knock stalks around so they are twisted and harvesters cannot cut through them.

Higher mean temperatures are only part of the story as higher low temperatures can be damaging. Some food crops such as pit fruits – peaches, plums and apricots – require a certain number of cold hours, typically 450 hours below 44° F. If the weather stays warm over winter and receives too few cold hours, the tree buds out in the spring but the buds fall off and the tree fails to set fruit. A recent study by UC David reported that chill hours have decreased 30% in central California, which has forced fruit orchards north.[187] A fruit orchard takes five years to grow trees that produce fruit, so tree farmers cannot just change to a different crop the following year.

Nobel-prize-winning physicist Steven Chu said in his first interview as Energy Secretary that California's farms and vineyards could vanish by the end of the century and its major cities could be in jeopardy if Americans do not act to slow the advance of global warming.[188] He also predicted that 90% of the Sierra Nevada snow pack, on which California cities and agriculture depend, would be gone by the end of the century.

Climate scenarios for 2020 project that Mexico will lose more than one million acres of maize production to hotter temperatures.[189] Corn production in the U.S. will also be forced to shift north. Temperatures in Vermont may be too warm to produce maple syrup, depriving farmers of their livelihoods.

David Battisti and Rosamond Naylor used data from 23 global climate models contributing to the Intergovernmental Panel on Climate Change's 2007 scientific synthesis to show with a greater than 90% probability that growing season temperatures in the tropics and subtropics by the end of the 21st century will exceed the most extreme seasonal temperatures recorded from 1900 to 2006. In temperate regions, the hottest seasons on record will represent the future norm in many locations. More than 3 billion people live in the tropics and subtropics, many of whom live on under $2 per day and depend primarily on agriculture for their livelihoods.[190] Their analysis portends catastrophic loss of food supplies. [191] The tropics and subtropics are likely to experience the worst food loss.

Additional heat will compound food insecurity caused by variable rainfall and will increase the incidence of agricultural droughts caused by accelerated evaporation from soils, transpiration from plants, low soil moisture and high rates of water runoff from hard pan soils when it rains. Excess heat causes virga – rain that evaporates before it hits the ground. Virga frustrates farmers when fields are dry.

Table 6.1 Climate Change Impacts on the Food Supply

Event	Impacts
Heat	Increased temperatures causes heat stress in food crops which can significantly diminish their productivity and lead to plant death and crop failure.
Hot winds	Increased temperatures and dry winds evaporate soil moisture and increase the need for freshwater irrigation.
Water scarcity	Water, the critical resource for sustainable food production, has passed its tipping point as global warming causes food crops to need more water but water in many growing areas water sources have been degraded, depleted or diverted.
Rising sea levels	Oceans will consume millions of acres of prime cropland on coasts and river deltas and tidal and storm surges will destroy millions of acres of cropland from sea salt invasion.
Ocean acidity	Dissolved CO_2 in the oceans diminish fisheries, destroy shellfish and dissolve coral reefs that protect coasts and estuaries.
Higher ocean surface temperatures	Heat creates the energy that intensifies storms, hurricanes and typhoons. Heat also changes the rainfall patterns and leads to drought and severe forest fires experienced in the western U.S.
Extended spring and fall	Spring is starting a week earlier and fall lasts an extra week, enabling pest vectors – bugs, fungi, molds, mildews, viruses and weeds – to multiply

	earlier and sometimes survive the winter.
Rain patterns	Shifts in rain patterns will cause huge losses of cropland that lack the infrastructure for irrigation.
Wildfires	Range lands and forests are especially vulnerable to heat, drought and winds that drive catastrophic wildfires such as those in California in 2008 and Victoria, Australia in 2009.
Loss of snow pack and glaciers	Snow packs are down 50% which means faster run off and heavy flooding in the spring. Reservoirs, creeks and rivers may be only half full when irrigation is needed later in the growing season. Melting snow packs and glaciers mean less river water for irrigation and human use.
Blowing dust	While the U.S. Mid-West experienced severe flooding in 2008, Texas and Oklahoma lost millions of acres of crops to drought and blowing dust. Dust decimates crops, amplifies drought by removing soil moisture and erodes thin topsoil.

Australia experienced its worst drought for more than a century in 2007 and saw its wheat crop shrink by 60%. China's grain harvest has fallen by 10% over the past seven years. India will soon overtake China in population and was food independent for a decade. Now, due to drought, salt invasion and failing freshwater supplies, India must import millions of tons of grain to sustain its population.

Heat creates additional drag on food supplies because higher temperatures mean more crops spoil in the field and, after harvest, before they arrive at food processors. Classic, cheap food grain storage in elevators and warehouses becomes less effective due to spoilage as well as higher rates of pest infestation.

In western North America, mean annual temperatures have increased at a rate of 0.7° F a decade and 1° F a decade at higher elevations which is causing widespread tree death. The tree death rate doubled in the Pacific Northwest in 17-year period and has doubled every 29 years in the U.S. interior.[192] Trees provide a natural CO_2 sink but when dead trees decompose, they release CO_2 into the atmosphere. Trees that are not dying are less healthy and more susceptible to pest invasion and wildfires. A Canadian Forest Service study found that the beetle outbreak in British Columbia has done so much damage that soon the forest will release more CO_2 than it absorbs.

Global climate change brings so many harsh variables that diminish food production that no single strategy can possibly address the diverse impacts. Transgenic seeds may offer modest relief in the sense that plants may be able to germinate in drier and saltier soils. However, little is known about the genetically modified seeds because they are still several years away and have not been tested in multiple growing situations. Transgenic seeds are often more vulnerable to pests, weeds and disease, which may offset other benefits. Many farmers cannot afford transgenic and others find their social belief systems forbid the use of genetically modified organisms. Transgenic seeds will simply magnify social injustice, the difference between rich and poor farmers, communities and countries.

Long supply chains

Fossil agriculture depends on more than 30 fossil inputs that must be affordable and available to farmers and their crops just on time. While a battle may be lost for the want of a nail, a crop may be lost for the want of a single element. The most critical fossil inputs, diesel fuel, fertilizers, fungicides, herbicides and pesticides have long and precarious supply chains. Long supply chains are not only expensive but risky because each fossil resource may be interrupted by supply-chain disruptions from whether, politics, war, terrorists, prices, transportation costs or other factors. A small supply disruption to fossil inputs could cause a major crop failure. Even a modest crop failure may ignite market forces to create an economic firestorm that leads to a resource run and possibly a full food cascade.

Social and political impacts

Sustaining food supplies depends not only on food production but the demographics of consumption, farmers and government policies. Growing food crops requires considerable physical labor and stable political policies that support food production.

New consumers. Americans have doubled their per capita meat consumption since 1990 and people around the world desire to emulate the consumption patterns of the rich.[193] The added demand for animal fodder substantially reduces available food grains. Meat consumption uses huge amounts of water since a pound of edible beef consumes 20 times more freshwater for irrigation than a pound of food grain.

By 2010, China will have twice the number of new consumers as the U.S., 600 million, who will demand higher value foods that create a high water exchange. China already leads the world in production and consumption of meat, primarily pork. Chinese per capita meat and milk consumption have doubled in the last 20 years.[194] China's increasingly high demand for grains combined with its distressed ecology and water problems, nearly all the lakes around Beijing have gone dry, will dramatically reduce world grain supplies. If a drought or food failure occurs, China could buy the entire U.S. grain output for a fraction of its positive U.S. trade balance.

Farmer succession. Over 40% of U.S. farmers are more than 55 years of age which jeopardizes future production unless replacements are found.[195] Farmer succession creates a very difficult problem because fewer young Americans choose to take the physical, psychological and financial risks associated with agriculture. Agricultural Land Grant Colleges are shifting from production to agribusiness because there are too few students who want to study farming. Young men who decide to farm find it difficult to attract a spouse who is willing to live with the isolation associated with rural life.

Globally, other factors are driving farmers to other professions. In India, a farmer makes about $1 a day. A factory worker makes $2 a day and an educated worker may make $5 to $20 a day. As more children receive education and they see other viable career opportunities, fewer choose farming.

Farmer health. HIV AIDS, malaria, malnutrition and other diseases limit many needed people from contributing to farm production. Disease vectors such as HIV AIDS tend to attack exactly the population needed for agricultural labor – young men. Failing sufficient labor, crop production diminishes or disappears.

Use of fertilizers, herbicides and pesticides has led to the death and disability of many farmers and their families, especially in developing countries. Farmers who have difficulty reading may adopt a "more is better" mantra and over apply agricultural chemicals or fail to use protective gear. The disabilities and deaths from agricultural chemical poisoning for farmers, their families and animals in several countries, including India, have been catastrophic.

Crop insurance. Farmers in the U.S. buy subsidized crop insurance to cover part of their crop, typically about 60%, in the event of crop failure or weather damage. Premiums are experience rated. In areas where crops have had damage from hurricanes, floods or wildfires, such as Louisiana and Florida, farmers cannot afford insurance. As more heat, drought and weather events occur, more farmers will be forced to leave their land.

Land redistribution. Political leaders may redistribute cropland to non-farmers who waste agricultural inputs and produce few crops. Government policies toward land redistribution, food prices and agricultural financing may undermine food production.

Food policies. Government controls on food production, distribution and pricing may make some crops impractical. Some countries, especially the U.S., have dealt a crushing blow to many farmers from subsidized foodgrains that are dumped on local markets as "food aid." Farmers cannot compete with subsidized U.S. grain and are forced off their farms. Canada, Mexico and several other countries have suits before the United Nations to stop the U.S. practice of dumping subsidized commodities.

Agricultural financing. Government or private sector limitations on agricultural loans for seeds and fertilizer may prohibit many farmers from planting productive food crops. Millions of farmers are in debt because the cost of fossil inputs keeps rising. These leveraged farmers are only one crop failure away from bankruptcy.

Crop subsidies. Food imports from countries such as the U.S. that are heavily subsidized undermine local food production because farmers cannot grow food profitably at the subsidized price. American farmers have received $177 billion in subsidies over the last 12 years while several countries have become dependent on U.S. food aid and have lost their ability to produce local food crops.[196]

Subsidized U.S. crops create an unsustainable dependency. Most of Haiti's farmers went bankrupt decades ago from U.S. food aid that dumped grains on Haiti's economy at prices lower than farmers could produce. Now Haiti is dependent on subsidized U.S. grain. Similarly, over 1.5 million Mexican farmers were forced to leave their land because they could not compete with subsidized U.S. corn.[197] Many of these farmers added their feet to the flow of illegal immigrants to the U.S. from Mexico.

British International Development Secretary Douglas Alexander said "It's unacceptable that rich countries still subsidize farming at $1 billion a day, costing poor farmers in developing countries $100 billion a year in lost income."[198] In spite of world opinion, the U.S. Farm Bill passed Congress in 2008 with nearly all the subsidies intact, especially for corn and ethanol.

Subsidized water and power. Many U.S. farmers receive enormous benefit from subsidized water, water distribution and the power to pump water. Subsidizing water creates excessive water waste. Farmers in California pay only 2% of the water cost for consumers in San Francisco. When water and power are provided nearly free to farmers, they have no incentive to protect the resource. The total taxpayer cost of crop, water, water distribution and power subsidies are incredibly expensive and may be worth over twice as much as the crops produced. Unfortunately, U.S. taxpayers have no knowledge of the true cost of commodity production because they are hidden

112

behind very, very expensive subsidies. Not only are these costs unsustainable but they mask the true social, economic and ecological costs of food production.

U.S. taxpayers would stop subsidies for irrigated corn used for ethanol if they understood that one gallon of ethanol consumed 3,000 gallons of freshwater.[199] The subsidy chain for ethanol includes the corn, irrigation water, power to pump irrigation water, 51 cents a gallon for refining ethanol a plus a litany of state and local subsidies for ethanol refineries.

Political instability. Political instability may make agricultural inputs insecure due to high prices, distribution failure or government controls on prices, tariffs and exports. Farmers lose their crops when political decisions thousands of miles away limit their access to agricultural inputs.

Political strife. War may conscript young men and take them out of the farm labor pool. Threat of physical harm may make working in fields too risky for farmers and cause crop loss. Some farmers cannot return to their fields because military ordinance litters their fields, especially land mines.

Monocropping

Roughly 75% of the world's food now comes from seven crops: wheat, rice, corn, potato, barley, cassava and sorghum. Farmers plant seeds from only narrow strains of each crop selected for efficiency in producing the most food the fastest in the minimum space. American's used to eat more than 15,000 varieties of apples but now less than 150 are commonly grown. The same winnowing of species applies to all food crops. Today, 99% of turkeys eaten in the U.S. come from a single breed, the Broad-Breasted White that has been bred for meat production but can no longer naturally reproduce. More than 80% of dairy cows are Holsteins and 75% of pigs come from just three breeds.[200]

Industrial agriculture maximizes production by focusing on a few crops. Over 10 million plant and animal species have been identified but more than 80% of the world's cropland is dominated by just 10 annual cereal grains, legumes and oilseeds. About 90% of the world's meat comes from only eight species of livestock. Wheat, rice and maize cover half the world's cropland and supply 80% of the total food produced worldwide.

Over 82% of U.S. farmland in 2006 was planted with the big four crops: corn, wheat, hay and soybeans. Corn subsidies motivated farmers who might have grown alternative crops to join the monocropping majority. Repeat plantings enable pests to multiply their numbers. Lack of biodiversity puts the entire production system at risk from a single vector that may be known or unknown such as:

- **Emerald Ash borer** has produced massive tree death throughout North America.
- **Water mould, fungus** which caused the Irish Potato Famine in 1845 and led to the starvation of nearly 1 million people, 12% of the Irish population.
- **Bacteria** which caused the U.S. Southern Corn Blight epidemic in the 1970s and reduced corn production to a fraction of normal in affected areas.
- **Micro-fungus** which killed most of the great Elm trees in the Midwest during the 1990s.
- **Sudden oak death** which kills many thousands of oaks annually in the West.
- **Bark beetles** which have devastated millions of forest acres in Alaska and the West.
- **West Nile virus** which kills millions of birds and dozens of people annually.

The Mediterranean fruit fly infested California in 1982, threatening billions of dollars worth of crops. Helicopter fleets sprayed pesticides such as Malathion at night and the National Guard set up highway checkpoints and collected many tons of local fruit. The state also released millions of sterile male flies and the pest was moderated.

In 2009, a caterpillar plague destroyed food supplies across West Africa where trillions of black, hairy larvae devoured plants, fouled wells with their feces and were so dense they drove farmers from their fields. This previously unknown species shares some similarities with army worms but has one tiny adaption; it climbs trees. Tree climbing may seem like a stupid pet trick but when farmers spray insecticides, the Liberian caterpillar stops eating the wheat, marches out of the field and climbs a tree where it is safe. The caterpillars then morph into butterflies that can fly for hundreds of miles and infect new fields. Aerial spraying is not an option because aerial application would contaminate water supplies.[201]

A soybean disease called the Asian blight affected wide regions of U.S. soybeans in 2008 and lead to substantial crop losses.[202] Phytophthora, a plant similar to fungi, attacks hundreds of different plant species including many crops, causing tens of billions of dollars in damage per year.[203] A single vector can decimate the entire production for large growing regions which makes monocropping a risky proposition.

Global warming intensifies problems with monocropping because more heat causes:

- Higher summer temperatures that accelerate growth of nearly every vector, including weeds, weevils and worms.
- Weaker plants from heat stress that increases susceptibility to disease and predators.
- More temperate winters that enables pests to survive over the winter which gives them the opportunity for exponentially faster growth with the new crop.
- Earlier spring, giving pests an earlier start time for propagation.

Monocropping creates severe dangers and damage due to lack of biodiversity, nutrient depletion of land and overuse of fertilizers, pesticides and herbicides. Monocropping also accelerates systemic soil erosion adding pollution to streams, rivers, lakes and well-water.

Technology rescue?

Sustainable food production is not possible with traditional land-based crops unless major breakthroughs occur for food crops in several areas:

- **Energy light** – requires minimal energy for food production.

- **Fossil light** – requires minimal fossil fuels, elements, metals and other fossil resources.
- **High productivity** – grows high yields of food on small acreage that diminishes demand for cropland.
- **Drought tolerance** – germinates and produces with low water requirements.
- **Salt tolerance** – thrives and produces in brine water and in soil invaded by irrigation salts or sea salt.
- **Weather tolerance** – can withstand temperature spikes, heat waves and fierce storms.
- **Low tillage** – requires modest tilling.
- **Perennial crops** – food grains and legumes that do not have to be replanted every year
- **Fertilizer light** – grows effectively with little added fertilizer.
- **Pest defense** – modest or no need for herbicides or pesticides.
- **Harvest waste** – harvestable with minimal waste.

Meaningful breakthroughs in these areas will take decades and cost significantly more than countries are willing to spend. Most these areas lack even a theoretical model. Therefore, a non-traditional food source is needed that meets SAFE production constraints.

The combination of forces and factors from population expansion, global climate change, long supply chains and social, health and political impacts challenge the sustainability of food production. In spite of the need for more food voiced by world leaders, traditional agriculture may not be able to deliver even current levels of food production.

SAFE production

Solving the social and political challenges requires a food source that can be produced with the following constraints. SAFE production does nothing to change population expansion except remove the motivation for poor farmers to have lots of children as insurance to work the fields.

Table 6.2 External Challenges

	Constraints for External Challenges
Social equity	People with the most need have access to affordable food inputs.
Global warming	Produces effectively with all the impacts of climate change.
Supply chain	Enables local production with a short supply chain.
Fodder	Produces high protein fodder for animals.
Health	Provides medicines for much needed illnesses including HIV / AIDS.
Physical labor	Diminishes physical labor needed to produce food and energy.
Risk	Lowers physical, health, mental and financial risks.
Subsidies	Stops ecologically destructive subsidies and invests in sustainable food and energy production.
Politics	Enables local production without political interference by a country's leaders.
Monocropping	Ends monocropping with a diverse set of food production alternatives.
Poisons	No or minimal pesticides, herbicides or fungicides.

Meeting these constraints with SAFE production will solve a set of the most critical issues of the 21st century.

Chapter 7. The Tiny Plant that Saved our Planet

The test of our progress is not whether we add more to the abundance of those who have much; it is whether we provide enough for those who have too little.

– Franklin D. Roosevelt

Algae's first Earth rescue was spectacular, systemic and slow – really, really slow.

Algae saved the Earth by employing green solar, capturing the sun's energy through photosynthesis to absorb CO_2, split the molecule to make hydrocarbon plant biomass while releasing pure oxygen into the atmosphere. Moving at the incredibly slow rate of one tiny molecule at a time, algae changed the incredibly harsh carbon dioxide atmosphere that could not sustain life to an oxygen atmosphere that supported life. Algae's atmospheric transformation supported the development of other water plants, fish, insects, amphibians, reptiles, land plants and animals.

About 3.5 billion years ago, the Earth's atmosphere was a whirl of inhospitable, fierce electrical storms. Driving winds and searing heat spiked with lightning abraded an atmosphere of CO_2, dust and sand across the Earth's surface. Lack of atmospheric cloud protection caused huge swings in day and night temperatures that intermittently froze and evaporated pools of water.

Abiogenesis, the study of how life on earth emerged, uses a primordial soup theory that suggests that the chemical conditions on Earth created the building blocks of life. While debate continues on exactly how the first life was synthesized, fossil records suggest one of the first plants was the size of a nano particle, cyanobacteria – blue-green algae.

The brutal conditions on Earth meant the first algae cells had to evolve and re-evolve many times as their microenvironments crashed with winds, severe heat followed by freezes and meteor showers of super-heated rocks. Algae displayed incredible persistence and developed a wide variety of defense mechanisms that allowed the plants to survive and to propagate.

Nature's oldest culture, green solar, converted water and CO_2 into chemical energy in plant bonds by using the sun's energy. Each molecule of CO_2 sequestered yielded a molecule of glucose, plant sugar, and gave off six molecules of pure O_2 to the atmosphere.

Figure 7.1. Algae Converts Carbon Dioxide to Oxygen and puts the Carbon into High-Energy Plant Bonds

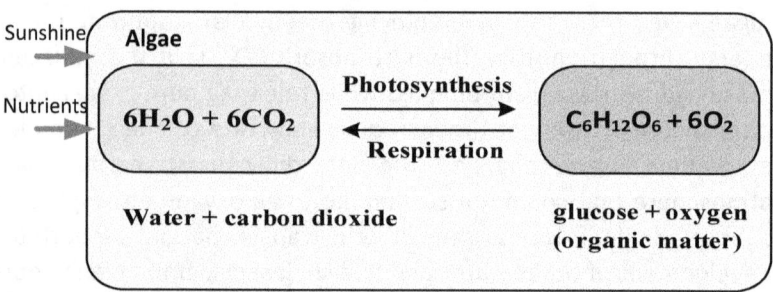

Algae's second save

After the atmospheric transformation, the Earth was an empty plate – no food. Algae solved this predicament by becoming the first rung on the food chain. Most the organisms that evolved later either ate algae directly or consumed higher tropic organisms that fed on algae.

Service as the lowest rung on the food chain meant algae had many predators so it developed hundreds of smart survival strategies including the ability to grow faster than any other plant on Earth. The many herbivores that fed on algae could eat many but not all of the fast growing plants. Algae may have been the first free lunch because it developed the capacity to double its biomass before midday.

Strategic adaptions

Algae evolved in millions of independent moist environments which created over 75,000 known species and possibly 10 million total species.[204] By contrast, insects and flowers each have about 400,000 species. Each algal species may express itself in multiple strains which make the choice of which algae to grow for food or energy practically unlimited. Each strain has its own nutrient profile and most favored growing conditions. Algae adapt quickly to changing conditions and a pure species culture may change its biomass composition in one morning on a hot day to adapt to increased solar energy and heat.

Microalgae are primitive organism with a simple cellular structure. Their large surface to volume body ratio gives each cell the capacity for a large uptake of nutrients. Algae typically live in moist environments where they are suspended in water which gives cells access to water, CO_2 and nutrients. When sufficient nutrients are available, algae's simple structure enables growth rates that are a high multiple of more complex land plants.

One of the secrets to algae's high growth velocity is the ability to propagate sexually or asexually. When conditions are good, algae reproduces sexually, enabling DNA transfer and evolution. When conditions become difficult, algae can propagate asexually using spores, cellular division or other strategies.

Various algae species have adapted to the extremes of cold and heat and may be found under the polar ice caps, under Tibetan glaciers and on and in the rocks of the Sahara desert. Each algal species grows best in a certain temperature range but many species are relatively temperature tolerant. Since algae have the ability to go dormant when their growing environment degrades, temperature spikes are not as deadly for algae as they are for land plants.

Algae evolved another defense mechanism perfect for a plant with many predators; hard cell walls. When a dinosaur ate algae, the animal's stomach could break down only about 30% of the algal cells. While passing through the gut, the surviving algae cells were in their favorite environment – warm, wet and nutrient rich – so what algae do best; grow and propagate. Well-nourished algal cells would then pass through the dino gut and be pooped into a warm, moist nitrogen rich environment – perfect for more algal growth.

Algae blooms were common in ancient oceans, lakes and ponds. Many species of algae are so tiny they are visible only under a microscope. However, algae may group, bunch, cluster or grow in formations that are visible and edible. Marine algae called seaweeds or macroalgae often grow into forms that have the appearance and size of land plants with pseudo roots, trunks and leaves. This parallel evolution enables marine algae to grow to sizes similar to trees. Algae provided a variety of bright colors for the oceans of antiquity and far more biomass than herbivores could eat.

In order to capture more solar energy so it could grow faster, algae embedded many pigments in its cells to capture light from multiple areas of the color spectrum. The colors of land plants including the rainbow of colors in flowers and vegetables come from algal pigments. Algal pigments provide color for many aquatic organisms including the beautiful pink color in wild salmon. Salmon farmers feed their fish algal pigments to obtain the desired color in their salmon.

The harsh conditions under which algae evolved over billions of years forced the plant to try millions of survival strategies. The useful strategies may still be found in many species. For example, land plants die when water or nutrients are not available or storms destroy fields. Algae developed a dormancy strategy where the plant just takes a break when conditions crash. As a consequence, a handful of dirt may harbor 100 algal species that become viable with moisture. Break open a rock and viable algae cells will probably be inside. Some species adapted to grow in forests, under polar ice, in the dirt and rocks under glaciers and others in the extreme heat of deserts. Algae can be found all over the planet growing in symbiosis with lichens, corals and sponges. Multiple algal species have adapted to every known microclimate on Earth.

A related adaption occurred around algae's nemesis; nutrient limitation. Many algae growing environments held standing water which quickly becomes nutrient limited because the plant grows so fast. Algae developed a set of strategies for dealing with this common problem:

1. Stop growing and go into dormancy until conditions reset.

2. Stop propagating but continue to store energy – often in the form of lipids.

3. Adapt to the nutrient limitation by absorbing organic nutrients for energy if inorganic nutrients were not available.

4. Produce lipids to create buoyancy which lifts the cell towards the top of the water column where there is light.

5. Grow flagella, a long whip-like tail that moves the alga cell to the needed nutrient. (Flagella movement is a very slow process, on the order of an inch an hour.) Some species can also grow eye buds that can detect the direction of light.

Algal adaption strategies create both blessings and curses for green solar farmers. The millions of adaptations benefit biomass producers by providing a wide array of choices for biomass composition. The downside of flexibility is that algae adapt with such velocity and complexity, systems can become unstable quickly. In order to create SAFE production, this wild plant must be tamed, domesticated and cultivated. Laboratory production has been optimized but outdoor cultivated algal production systems, CAPS, have not.

When conditions are good, algae grow and develop rapidly. When algae are stressed by nutrient limitations, they slow growth and immediately begin storing fats to ensure their survival. Green solar energy producers take advantage of this adaption and starve algae of nitrogen so it will double the production of lipids that are useful for liquid transportation fuels.

Algae's third save

Algae hold promise to save the earth a third time by doing what land plants cannot do – recover and recycle fossil nutrients lost in human, animal and industrial waste streams. Land-based food crops could not recover and recycle fossil nutrients even if waste streams were used for irrigation for two reasons:

1. The dilute nutrient solutions prevent plants from absorbing sufficient amounts of each nutrient. The plant simply cannot process enough water to get enough nutrient energy.
2. The salt ions in waste streams clog the plant's root system, starving the plant of nutrients and causing plant death.

Algae require the same fertilizers as land plants but do not have to rely on the expense and ecological problems associated with mining fossil resources. One of algae's adaptive strategies, bioaccumulation, enables the plant to concentrate certain elements up to 1000 times their ambient levels. Algae grown in wastewater, brine or sea water can accumulate organic and inorganic nutrients from very low dilutions. Algae do not present the problem of salt ions clogging the root systems because algae have no roots. Algae evolved in salt water and many freshwater species are salt tolerant.

One strong application for algal production will be the capture of fossil nutrients in wastewater and oceans through bioaccumulation that may be recovered in solid algal biomass. When the salts and heavy metals are separated from the biomass, algae can provide the rich organic fertilizer and soil conditioner to support traditional land-based agriculture.

Green solar energy

Green solar, cultivated algae, captures the sun's energy in water-based plants and employs photosynthesis to store solar energy in carbon molecules in algal plant bonds. Algae may be 60% by weight food energy for people, nutrient energy for plants or fuel energy for vehicles. Algae have no reason to put energy into roots, trunks and leaves because they grow in water where water supports the plant in their natural settings. Algae are energy positive because the energy cost of algaculture and oil extraction are minimal while the energy produced is higher than ethanol or gasoline.

Green solar has no growing season since half of the high protein biomass may be harvested daily. Most species have a preferred production period which typically corresponds to land plants but they can be cultivated year round. Algaculture produces considerable O_2 while sequestering CO_2 so it provides a positive air footprint.

Green solar provides a portable energy source and grows biomass with solar energy stored in forms that may be used for a variety of purposes:

- **People** – organic protein in food
- **Animals** – organic protein in fodder
- **Fowl** – natural protein for birds
- **Fish** – natural protein in fish feed
- **Land plants** – organic nitrogen fertilizer
- **Fire** – high energy algal oil for cooking and heating
- **Cars** – carbohydrates refined to gasoline for transportation
- **Trucks and tractors** – high energy clean, green diesel
- **Trains, boats and ships** – high energy clean diesel
- **Planes** – high energy, clean aviation gas and jet fuel

Algae serve as food for 100 times more organisms than any other plant. The many species also means the plant displays a wide variety of nutrient profiles for its many hungry consumers.

Green solar creates energy similarly to land plants by consuming CO_2 and releasing oxygen in the same process as the brown solar energy in trees and bushes and grey solar in food grains. Land plants evolved from algae and had to make substantial compromises to survive on land. They had to grow roots and put much of their energy into non-food structures such as trunks, stems, leaves and reproductive organs in order to survive and grow in an open air environment. This costly overhead creates a huge production drag that slows growth and food productivity compared with water-based plants.

Land plants require an entire growing season, effectively a whole year, to grow the first ounce of biomass that contains protein for food or lipids or carbohydrates for biofuels. Green solar is 30 – 100 times more productive than land plants because algae produce protein and biofuel in the same culture and produce high-energy biomass daily, every day the sun shines

Cultivated algal production systems, CAPS are designed to maximize the plant cells' exposure to solar energy which drives photosynthesis. Some green solar systems are open ponds which, like traditional fields, absorb solar energy only from above. Open ponds employ some form of mixing to maximize solar exposure because algae grow so fast that the top plants shade those only two inches below. Algal cells that do not receive sunlight stop growing so mixing serves to maximize culture growth.

Closed CAPS may look like thin aquariums and may absorb solar energy from above and the side facing the sun. CAPS productivity is directly related to the amount of solar exposure, similar to photovoltaic cells. Consequently, optimal growing systems may be vertical, lateral or tubular depending on the setting and the production goals. CAPS may be placed on rooftops, balconies, sides of buildings, fences, backyards or anywhere on Earth that receives solar exposure and grow harvestable biomass each day the sun shines. In extremely cloudy days and at night, the culture rests in the respiration phase of photosynthesis.

While water plants create the great majority of green solar energy there is a fascinating exception. A sea slug was recently discovered that appears to be a plant-animal hybrid. The slug's cell tissues have the light harvesting components from plant cells and a key gene that enables photosynthesis. This sea slug grows not by eating normal organic food but by gaining its energy from the sun. Most algal species are similar hybrids in the sense that they are primarily photosynthetic and get their energy from the sun and inorganic nutrients. However, when their inorganic nutrients become limited, many species can use organic sources for energy.

While algae create biomass prolifically, it also captures and stores carbon, lots of carbon.

Carbon capture

Much of the green solar energy algae captured in algae plant bonds fell to the bottom of the water column and was sequestered in the muck. Green solar sequesters 1.8 pounds of CO_2 in every pound of algae. The biomass provided energy via food for plankton, fish, reptiles, birds and for larger animals such as dinosaurs and whales. These creatures ate algae or algal feeders and stored the green solar energy in their carbon biomass. When these creatures died, their organic remains also entered the muck. Over hundreds of millions of years, the Earth's geology pushed the muck deep in the earth under high temperatures and pressure where it fossilized and was stored for eons.

When man discovered this long sequestered deep carbon it was no longer green because fossilization had turned it into black petroleum, coal, shale and natural gas. This fossilized black solar energy turned out to be an ideal, transportable liquid transportation fuel. Coal provided a low-cost fuel source for heating steam turbines that ran ships, trains and electrical power plants. Consumption of fossilized carbon accelerated as people found applications for fossil fuels in agriculture, industry, transportation and electrical power generation.

Unfortunately, reintroducing fossil carbon to the atmosphere created a heavy carbon load to the atmosphere and undermined the critical sequestration work done by algae over billions of years. In only 150 years, human consumption of one trillion barrels of fossilized solar energy in fossil fuels pumped trillions of tons of carbon into the atmosphere.

One gallon of gasoline weighs only 6.3 pounds but produces 20 pounds of CO_2 because burning hydrocarbons splits the carbon atoms from their hydrogen bonds, creating water, while the carbon combines with oxygen in the air to produce CO_2. Hydrogen is light, with a mass of one, while oxygen has a mass of 16, so gasoline creates about three times its weight in carbon load to the atmosphere. The release of fossilized carbon created profound effects on the environment including climate change that threatens human societies in terms of health, food, water, transportation and consumer goods and life-support systems.

Carbon recapture

Algae provide the only viable solution re-sequestering atmospheric CO_2 because green solar requires only plentiful and cheap resources and little or no fossil energy. Mechanical solutions to remove CO_2 from the air are far too expensive in terms of dollars and the fossil energy needed. Other approaches such as growing more plants or trees consume large amounts of fossil fuels and provide very modest carbon capture. Forests and grasslands provide only temporary carbon sequestration because when the trees and grasses die, their stored CO_2 is re-released to the atmosphere.

Green solar, nature's original solution to carbon capture, holds excellent potential to serve again to reclaim the Earth's atmosphere. This process, also known as algaculture and nanoculture, uses the sun's energy through photosynthesis to capture "top carbon," CO_2 in water or the air. Even though algae is the tiniest plant on earth, representing only 0.5% of the Earth's plant biomass, it creates about 60% of the Earth's oxygen — more than all the forests and fields combined. In the process of capturing carbon, algae also create harvestable green energy.

Carbon neutral energy

Green solar offers a sustainable, low cost and non-polluting source for creating carbon neutral biofuels in weeks instead of millions of years. Algae use abundant, low cost inputs to capture carbon and create energy. Sunshine is free, waste water for nutrients are often free and CO_2 is often surplus from manufacturing or power plants. Growing algae requires modest energy inputs to mix and circulate the water and for harvest and component extraction. This energy can be supplied by non-fossil sources such as solar panels, windmills or surplus algal oil.

The biomass of lipids (oils), protein and carbohydrates may be 20 to 50% oil. The oil may be pressed out of the biomass to create clean biodiesel (vegetable oil) that fuels diesel engines directly without modification. Algae costs about as much to produce per pound today as beef and consumers typically prefer to eat beef. However, large-scale algae farms that will come online in the next few years promise algae oil at a 20% discount to fossil diesel with neither the fossil CO_2 nor the black soot particulates. Algae also hold promise for food protein at a 40% discount to feed grains. The protein coproduct after algal oil extraction may be 50% of the remaining biomass.

Ecologically positive

Unlike fossil fuels, algal oil burns cleanly because it is the same as vegetable oil that is burned in many diesel engines today. Algal oil produces recycled CO_2 rather than adding to the fossil CO_2 load. Fossil fuels must be mined which creates a series of pollution problems including the long supply chain from the drill site to consumption. Biofuels such as corn add 2.5 tons of CO_2 per acre plus nitric oxides, particulates and smog. Biofuels from land crops also create significant pollution to soils, air and water.

Green solar can be produced in closed or semi-closed containers that are ecologically positive and non-polluting. Production can occur anywhere which enables a short and secure supply chain. Green solar farms do not need freshwater because algae thrive in waste, brine or salt water. The growing process gives off pure O_2 to the atmosphere and does not give off nitric oxides, soot or smog associated with land-based biofuels. Green solar uses no or few agricultural chemicals, herbicides or pesticides and culture water may be recycled so there is no air, soil or water pollution.

Health

Possibly the highest value algal coproducts are medicines, pharmaceuticals, vaccines and nutraceuticals. Algae are used to make capsules and tablets, stabilizing agents for stomachs, suppositories, radiology agents, anticoagulants and antiulcer lotions and creams. Fish get their healthful Omega 3 fatty acids and other beneficial characteristics from algae and are popular in the nutraceutical market. Several nutraceutical brands produce Omega 3's directly from the algae that are more environmentally sustainable and store less mercury than fish. Published research on algal medical applications include: infections, external wounds, pesticide poisoning, obesity, cancer, diabetes, hepatitis, pancreatitis, cataracts, constipation and allergies.

A large number of antibiotic compounds, many with novel structures, have been isolated and characterized in algae. Cyanobacteria have been able to produce antiviral, antineoplastic and other pharmacologically active compounds.[205] A variety of algal compounds are active against herpes virus, pneumonia, SARS and HIV. Some produce antitumor and antifungal compounds. Bioactive algae compounds are finding applications in both human and veterinary medicine and in agriculture. Antioxidants provide value for the food industry. Lipophilic antioxidants serve as food preservatives, preventing lipid peroxidation which causes food spoilage. Other applications include research tools or structural models for the development of new drugs.

Scientists have found algal species that contain compounds that inhibit the division of cancer cells grown in the laboratory. Additionally, a compound isolated from algae collected from oil platforms in the Gulf of Mexico has been shown to block cell division and enhance the activity of the cancer drug Taxol.[206]

Guoyao Wu and team at Texas A&M University found a potential obesity solution in dietary arginine supplementation that shifts nutrient energy to promote skeletal-muscle gain rather than fat.[207] Arginine extracted from algae seems to increase lean tissue and muscle growth while diminishing accumulation of fat. Arginine also reduces serum concentrations of branched-chain amino acids and may lead to insulin resistance in obesity.

Intensive R&D is focused on technologies that will enable algae to serve as medical protein factories. Recombinant human proteins expressed in plants tend to show the same activity as the original protein. Algal cultivation is faster, e.g. days versus months, and less costly than using higher plants or animals as biofactories. One of the major difficulties with using plant models is that the recombinant protein must be purified from the plant material and those activities takes time and energy. Proteins are easier to separate from single-celled organisms such as algae. Green algae fall into the category of Generally Recognized as Safe (GRAS), meaning they are safe to eat and offer a source for the delivery of therapeutic proteins.

Biomass components

Algal biomass contains three primary components proteins, lipids and carbohydrates. Various algae maximize different components. Some species offer more than 65% protein, others 70% lipids and others 90% carbohydrates. The lipids and fatty acids are found in membrane components, storage products, metabolites and sources of energy. Algal oil is similar to fish and vegetable oil and can serve as a fuel oil. Diesel trucks running on algal oil smell a bit like French fries because they are burning vegetable oil from a water-based plant.

Table 7.1 Algae Components

Component	Description
Protein	A food source, depending on the algal species, that may provide a better nutrition profile than land plants such as grains.
Lipids	Algal oil that may be used directly as cooking oil or for cooking fires. Like other vegetable oils, it is used directly as a fuel in diesel vehicles. Algal lipids may be processed into any fuel produced by crude oil.
Carbohydrates	Similar to land plants that may add value to foods, food ingredients or additives or be refined into ethanol, methanol and other fuels.

Land plants must invest extensive resources in a dense woody cellulosic structure – trunk, roots, leaves and stems. Algae waste no growing energy or biomass on cellulosic components because they do not need structure; they are supported by nature's womb – water. These single-celled organisms are so elementary in a genetic sense they have no specialized roots, trunks, leaves, stems or reproductive organs. These structural factors enable algae to offer far higher protein production than land plants per unit of biomass, Figure 6.2.

Figure 7.2. Energy Expended for Land plants vs. Water Plants

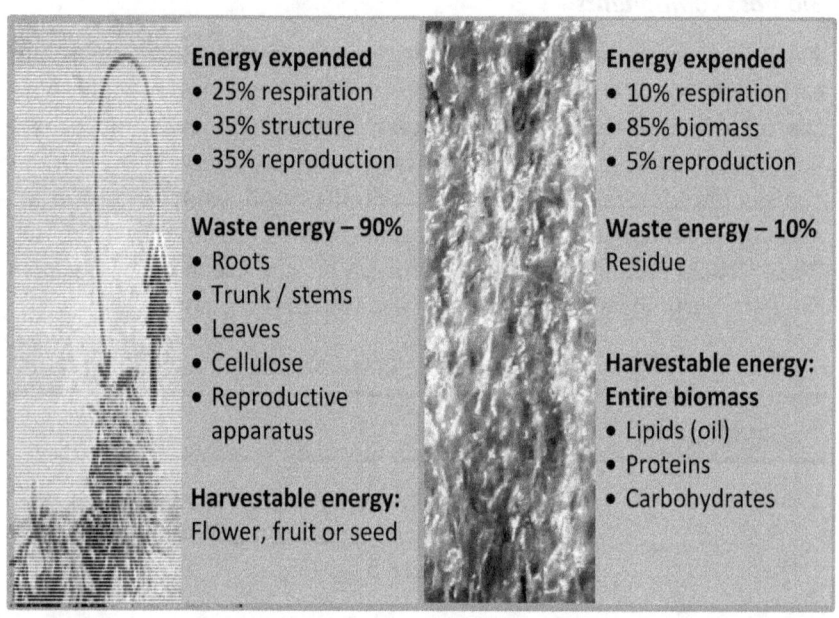

Energy expended
- 25% respiration
- 35% structure
- 35% reproduction

Waste energy – 90%
- Roots
- Trunk / stems
- Leaves
- Cellulose
- Reproductive apparatus

Harvestable energy:
Flower, fruit or seed

Energy expended
- 10% respiration
- 85% biomass
- 5% reproduction

Waste energy – 10%
Residue

Harvestable energy:
Entire biomass
- Lipids (oil)
- Proteins
- Carbohydrates

A modest exception to the cellulosic rule has recently been discovered by Patrick Martone and his team at Stanford University that found lignin in the cell walls of red algae that must endure the pounding coastal waves.[208] Lignin is a glue-like substance that helps fortify cell walls and is instrumental in the transport of water in many plants. The finding shows that algae developed cellulosic materials before land plants evolved from green algae.

Under normal growing conditions, most land plants will produce biomass that has similar composition – a nut stays a nut. Algae are far more adaptable and have the ability to change biomass composition based on growing conditions. Algal biomass composition may change with variations in available light, nutrients, culture acidity (pH) or temperature. Composition may change due to age or density of the algal culture. In dense cultures, algal cells tend to store more lipids (fats) or more carbohydrates, depending on the species. Less dense cultures tend to be more protein rich. Several other factors may cause algae to store more lipids such as limiting nutrients, especially CO_2 or nitrogen at critical time in the culture's growth.

Land plants such as corn produce short carbon chains that yield low energy fuels such as ethanol. Corn can be converted to ethanol that burns with less heat and provides only 64% of the MPG of gasoline. Some algal species offer medium and long hydrocarbon chains that can be converted to more powerful liquid transportation fuels such as JP-8, jet fuel and green diesel that may have 30 to 50% more energy per gallon than gasoline.

This extraordinary plant offers substantial value as a food and energy source but remains largely undomesticated.

Why have algae remained undiscovered?

The most common algae question is:

> If algae have so much potential, why has less than 1% of its potential been realized?

For decades, food and fuels were so cheap that there were no incentives for algaculture. Today, soy protein can be grown at about one tenth the cost of algal protein. Fossil fuels can be extracted and refined for about one fifth the cost of algal oil. Those numbers will flip with scaled commercial algaculture but those technologies have not yet been demonstrated commercially. Green solar is not currently cultivated in the U.S. for a variety of reasons.

Politics. Algae and other truly sustainable food and biofuel feedstocks have lost every battle to corn as politics by both political parties repeatedly trumped science.[209] The U.S. government made a political decision to back corn ethanol in the 1990s as America's "renewable" biofuel and eliminated federal funding for algae at government agencies like DOE and USDA and university and private research and labs. Without government funding, universities, the National Labs and public and private research institutions were forced to disband their algae laboratories and research because there were no grant monies available. Algae have no political action committees, no lobbyist or spokespersons.

Sex appeal. Algae do not have the same appeal as deep space exploration. NASA receives $17.3 billion for space exploration but produces neither an ounce of food nor a drop of fuel. NASA has sponsored some algal research, especially for extended space flight.

Subsidies. Algae receive no subsidies similar to corn, wheat and Big Oil. Algae receive no government incentives for building or operating algal production systems. Algal producers cannot find funding because their loans receive nether needed guarantees nor incentives similar to ethanol plants.

One of the major issues facing the algal industry is the lack of trained professors and graduate students to staff algae laboratories. The few algae labs currently operating the U.S. employ staffs that are predominately foreign trained. The 2007 Energy Security and Independence Act includes language promoting the use of other renewables such as algae for biofuels. Algae began receiving minute funding in 2007 and NREL has reestablished some research on algae. However, algal research still receives a fraction, less than 0.1% of the subsidies for corn ethanol.

Algae have remained undiscovered also due to the strong negative social attribution. On an algal belief survey, more than 92% of consumers responded with "dislike intensely." They dislike algae because they associate it with icky, stinky green slime.[210] Consumers seem to have a natural aversion toward something they cannot see because it is too small. Of course, consumers cannot see plant cells either but they are familiar with the form of traditional land plants.

Most consumers have near zero knowledge of algae yet they share an extremely strong negative perception. Consumers are unlikely to embrace a food with a green smelly slime legacy. Green solar provides an alternative to the term algae which creates a positive social attribution and positions algae correctly a form of solar energy.

A broad set of private equity firms are making investments in algal oil production for fuels but the risks for the first investors in this infant industry are very high. Breakthroughs are likely to enable cheap solar to produce electricity stored in batteries to power cars at an order of magnitude more cheaply than any biofuel. The same probability holds true for blue solar energy sources including wind, waves and geothermal. The energy opportunity for Green solar lies in the production of non-fossil, carbon neutral liquid transportation fuels as well as food, fodder, fine medicines and other coproducts.

Lack of government funding for algal research for sustainable food solutions creates a threat that breakthroughs will occur too late and traditional agriculture will fail to provide sufficient food. This highly probable outcome will lead to a global food cascade the starvation of millions. Failing public funding, private firms will lock up the fundamental pathways for algal production with intellectual property protections that prevent wide-spread adoption of green solar. Without public funding and open source technologies, scientists may discover viable solutions that may put an end to hunger but these innovations will be locked in corporate intellectual property safes and the people with the most need will have no access.

The primary obstacle holding abundant agriculture back is lack of public investment. The nascent algal industry exists as a series of vertical markets with each firm holding tight to its own intellectual property and sharing little about their scientific breakthroughs. Algal biomass conferences are filled with speakers making grand promises but showing almost no data because they have non-disclosure agreements with their employers. Only a very few people know how to grow algae effectively and most are hoarding their knowledge.

To solve this dilemma, http://GreenIndependence.org shares algal knowledge with open source technology available free to anyone on Earth. The site includes a sticky wiki that enables questions, discussions and research threads, research papers, project summaries, industry analysis and links to useful resources.

GreenIndpendence.org is dedicated to freedom from hunger, oil imports, ending the need for fossil fuels and energy and nutrient recovery from human and animal waste streams. The Green Algae Strategy Book Series content, including *Green Algae Strategy* and *Crash!* are available in color PDF for free download to students, teachers, scientists and others engaged in food and energy policy at http://GreenIndependence.org.

Chapter 8. Abundant Agriculture

If one way be better than another, that you may be sure is nature's way.

– Aristotle

Modern agriculture is not sustainable due to its heavy reliance on non-renewable fossil resources. Organic farming reduces reliance on fossil fuels somewhat by avoiding synthetic fertilizers and most herbicides and pesticides. Unfortunately, organic farmers are nearly as dependent on fossil resources as other farmers because they use considerable fossil fuel in farm machinery, especially for plowing under animal or green manure. Many organic farmers are reliant on supplemental fossil fertilizers such as phosphorus, sulfur and zinc that will be depleted in the lifetime of their children. Similarly, many farmers are dependent on fossil water which may be depleted in their lifetime.

Organic farming is not sustainable because growing enough green manure, vegetative wastes, for organic fertilizer would take several times more cropland than is available for growing food. Organic and modern farmers are equally at risk from the vulgarities of global climate change including especially temperature spikes, drought, fierce storms, sea and irrigation salt invasion, erosion, rising ocean levels and pest infestation. Organic farming often suffers from low productivity, pest and weed invasions while consuming as much freshwater as modern agriculture.

Abundant agriculture

Aristotle was correct in suggesting nature's way. Abundant agriculture makes use of Earth's oldest natural growing system, green solar or algaculture, and avoids the use of fossil resources. Green solar produces food and fuel using plentiful resources that will not run out – sunshine, waste, brine or ocean water and CO_2 – and is weather insensitive when grown in closed production systems. A sustainable and affordable food and energy (SAFE) production system that overcomes many of the disadvantages of traditional fossil agriculture offers hope for millions of hungry people. Abundant agriculture is practical because it leverages the robust characteristics of the Earth's first and simplest food.

Algae can rescue Earth again by leveraging its dominate behavior – capturing and storing solar energy and carbon to grow valuable biomass. We cannot afford to wait another 400 million years for nature to act for this rescue, so algae will need our help.

Table 8.1 Sustainable non-fossil Food – Abundant Agriculture

Challenge	How green solar differs from fossil agriculture
Renewable, affordable and sustainable	**Good cropland** – requires no cropland and may be grown in deserts, waste dumps, rooftops, parking lots, sides of buildings or wasteland.
	Freshwater– requires no freshwater and may be grown in wastewater, brine water or sea water.
	Fossil fuels – requires light energy for mixing water and harvesting which can be provided by solar, wind or the electrical grid.
High cost inputs	Avoids the cost of cropland, heavy labor, seeds water and fossil resources such as fertilizer.
	Uses abundant low-cost resources such as wasteland, wastewater, waste CO_2 and sunshine.

There is a cost in and building and operating a green solar system which we expect to be substantially lower than traditional agriculture.

Robust production	Closed or partially closed systems can produce food independent of geography or weather.
Growing season	Algae can double its biomass daily and may be harvested daily or every few days.
Pollution	Neither open ponds nor closed or partially closed systems pollute the local ecology. Algae give off no greenhouse gases and release pure oxygen.
Soil erosion	Algae do not grow in soil and therefore create no soil erosion or soil pollution.
Salt invasion	Algae have no roots so they are not only salt tolerant but actually prefer water that is slightly more saline than found in their natural setting.
Financial and physical risk	Food production in weeks rather than an entire growing season reduces financial risk.
	Physical risk reduction occurs from eliminating the need for heavy mechanical equipment and the use of poisonous agricultural chemicals.
Crop diversity	With more than 75,000 species from which to select protein composition and nutrient profiles, producers will be able to produce a broad array of different foods. Algae may also be fed to shellfish, finfish, poultry and other animals which expand food diversity exponentially.

Minimal herbicides and pesticides	Solar farming may use non-chemical methods for maintaining culture stability such as controlling parameters such as temperature and pH. Heating water in natural solar heaters can eliminate weed algae, bacteria and parasites from the culture.
Pollinators	Algae do not depend on pollination because the plant has no reproductive organs. Algae reproduce by cellular division, fragmentation or spores, none of which require pollination.

In closed production systems, algal biomass can be grown anywhere on Earth where the sun shines or artificial light is available. Therefore, green solar produces independent of altitude, latitude, longitude, climate and geology. Algae like heat so it naturally benefits from global warming. Protection in closed or shaded biofactories makes SAFE production independent of winds and weather.

The human labor required for fossil fuel production is significant whether oil is drilled or coal is mined. The work is physically demanding and risky for accidents and detrimental to health. Farming, mining and fishing have the highest rates for disability and death from accidents and exposure to chemicals. They also are high on the list of vocations that require physical strength and stamina.

The human labor required by green solar is more like gardening than mining or farming. Avoiding heavy equipment, huge engines, chemical and particulate pollutants removes most the risk for accidents and health. Some unanticipated issues are certain to arise in green solar production but of most the physical labor and health risks are removed from the energy and food production models.

Sequestering CO_2 in algal biomass is a coproduct of producing energy. While much of the CO_2 will be released in the green energy, it will displace not only the sequestered CO_2 from fossil fuels but avoid the other air, soil and water pollutants, including especially mercury, arsenic and black soot particulates.

Examination of production metrics suggests that every ton of algae produced for energy will net an absolute reduction of about 20% of algae's 1.82 tons of sequestered carbon. This occurs because the residual carbon not burned as a biofuel will be distributed as animal fodder or incorporated in soils as organic fertilizer. Algae captures more than its own weight in CO_2 because the lighter carbon atoms are made into biomass while algae releases the heavier O_2 molecules.

Green solar energy solutions

The key to green solar solutions will be total cost because change will occur only with the incentive to create energy faster, cheaper and easier. While production costs seem to be an obstacle currently, no one has experience with large scale green solar production. Conservative production models indicate green solar production costs are far less than drilling, mining and transporting fossil fuels.

Governments, businesses and leaders need to consider the total costs of various energy alternatives including:

- **Direct costs** – the total amount of money required per unit of energy production.
- **Human costs** – physical labor, physical risk and direct and indirect health risks. Mining coal, for example, puts not only miners at serious risk but all families, especially children, who live within the pollution plumes of the mine, the rail used for transport and the coal-fired plant.
- **Air pollution** – the extraordinary costs associated with global warming, greenhouse gases and black soot particulates.
- **Water pollution** – the consumption, pollution, security and ecological effects on fresh water.
- **Soil pollution and erosion** – mining displaces soils and pollutes with heavy metals such as mercury and arsenic. Massive growing of corn pollutes soils with agricultural chemicals and tilling creates wind and water erosion.
- **Indirect costs** – the externalities associated with fossil fuels are huge including loss of fossil resources for future generations, the health, food crop and weather effects of pollution plumes, overconsumption of fresh water and environmental degradation.

Green solar offers a set of positive water solutions.

Green solar water solutions

Neither algae nor any other technology can manufacture water. Desalinization provides at best a stopgap measure for cities desperate for drinking water. Desalinization is far too expensive in both costs and energy to use for crop irrigation. Algae offer two strategies for supporting sustainable water: displacement and replacement.

Table 8.2 Green Solar Water Solutions

Action	Description
Displacement	Green solar can clean waste or salt water thereby displacing freshwater that would otherwise require clean irrigation water.
Water savings	Algae grown in closed loop algaculture systems have minimal evaporation or water loss. The water used for growth can be recycled which saves both water and residual nutrients.
Brine, waste or salt water	Algae grow well in water that would kill land crops such as brine, waste or salt water. Algae have no roots so salt ions do not interfere with water uptake or transport.
Replacement	Algae replace polluted water with water that is sufficiently clean that it can be used for irrigation or with treatment, for human use.
Clean human wastewater	Algae assimilate nutrients from water treatment plants and return the water to a near-drinkable state.
Clean industrial wastewater	Algal bioremediation absorbs heavy metals, toxic compounds and poisons from industrial wastes. These elements and compounds can

be recovered from the algal biomass and may be sold or reused. Heavy metals are easier to separate from when concentrated in solids, algae biomass, than when they are in solution.

Clean agricultural wastewater	Conventional technologies for removing nitrogen and agricultural waste stream chemicals are very expensive and use considerable energy. Algae provide and inexpensive solution and require light energy that can be supplied with green fuels.
Remove pharmaceuticals from lakes	Discarded prescription and over-the-counter drugs have polluted lakes and well-water which can be cleaned with specific species of algae.
Industrial heated water	Many industrial manufacturing plants as well as coal-fired power plants and nuclear energy facilities use extensive fresh water for cooling. The industrial heat can be flued through algal ponds that absorb the heat which acts as a catalyst to speed green biomass production.

Algae also offer water and air pollutions solutions in that algae may be used to test, diagnose and monitor for a wide range of potential toxins and other contaminants. One of algae's defense strategies is to react in the presence of certain dangerous chemicals in very low dilutions. These reactions can serve as a monitoring system for air or water pollution.

An additional constraint to sustainable energy includes the need for no or minimal irrigation or fresh water to grow the biofuel. Green solar grown in open ponds loses nearly as much water to evaporation as was lost in irrigation of field crops in hot climates. Closed CAPS that recycle water may use only 0.001 as much fresh water per gallon of biofuel as irrigated corn or 0.01 as much as mining and burning coal. This minimal water requirement provides significant advantage for algal production for the many global cities, towns and villages that are water insecure.

Green solar for food

The cultivation of communities of single and multicelled organisms in green solar farms or gardens produce vibrant communities. Light makes life as these plants grow quickly and transform from invisible forests of nano sized cells to rich, dense biomass clouds or to larger organisms known as macroalgae or seaweeds.

Green solar production may flow into secondary containers where small filter feeders such as phytoplankton or brine shrimp harvest the algae by eating it. The secondary containers may include organisms of higher trophic levels such as shellfish or finfish that feed on algae directly or the phytoplankton or brine shrimp. The use of organisms that feed on algae provide several advantages including that feeders are easier to harvest since they are larger food sources that can be filtered with nets and these fish or shellfish may be eaten directly.

Many algae species have a good or neutral taste but people are often reluctant to eat something that is too small to see. Of course, the individual cells of land plants are not visible but they grow into a form that offers a common appearance. Harvested algae look similar to a flour or cornmeal. The algal biomass can be made into anything made from soy, corn, rice or wheat protein such as chips, tortillas, crepes, breads, nuts, energy or health bars, cakes, sauces, soups or tofu.

Initially green solar producers will probably produce food ingredients and nutrients that add value to traditional foods rather than eating the algae directly, Figure 7.2. Algae typically have nutrient profiles superior to land plants, especially important minerals and natural vitamins. Therefore food ingredients could enrich foods, improve health, extend shelf life and enhance texture, flavor, colors and aromas. Much of the fish oil sold today comes from algal producers in Israel. Fish are high in Omega-3's and other vitamins because they get their nutrients from algae. Fish may concentrate mercury if their environment is contaminated. Algae-based Omega-3's do not have mercury contamination because they are grown in a controlled CAPS.

Figure 8.2. How Algae is used for Food

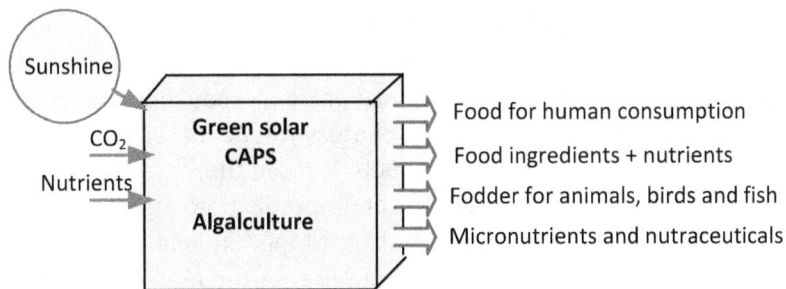

Growing algae as fodder for animals, birds or aquatic creatures will be popular in many settings because animal fodder requires lower levels of cleanliness than producing food for direct human consumption. Algae can provide high-protein fodder for animals such as cows, goats, sheep, hogs and poultry grown for human consumption. Locally grown algae also provide a coproduct in low cost yet rich organic fertilizer that increases the productivity of land-based crops.

Most medicines, vaccines in pharmaceuticals are grown today in animal or plant models. For example, harvesting a medicine in cow's milk takes considerable time to grow the grasses the cow eats and then to harvest the medicine from the milk. Production of medicines from algae could lower the production cost substantially because growth and harvest could occur in days rather than months. In addition, producing medicines in algae grown locally may be consumed directly in the algae. Direct consumption would eliminate the incredibly high costs of extraction, formulation, packaging, storage and transportation.

Selected algal species that are nitrogen fixing may be grown locally and placed on land crops such as rice to improve fertilization and food productivity. Research has shown rice productivity may increase 10 times with the addition of nitrogen fixing algae in flooded rice paddies. Other algal species grow symbiotically with land food plants and help them be more productive.

Even though algal cells are small, a single cell can produce more than a million offspring in one day. Under proper growing conditions, some species multiply their biomass two, three or even four times a day. It flourishes when the sun shines and rests at night or on cloudy days.

Algae's food value has been known for centuries and food potential for at least 100 years. Consider the annual protein production per acre for food grains, Figure 7.3. Annual protein production per acre of algae is calculated based on its photosynthetic capacity which is 30,000 pounds per acre but 15,000 is used here as a practical estimate.[211] In addition to higher protein production, algae provide a superior set of vitamins and minerals not found in land plants. Algae are not a full solution for malnutrition because the biomass is low on calories.

Figure 8.3. Algal Protein Production Potential – Pounds per Acre per Year

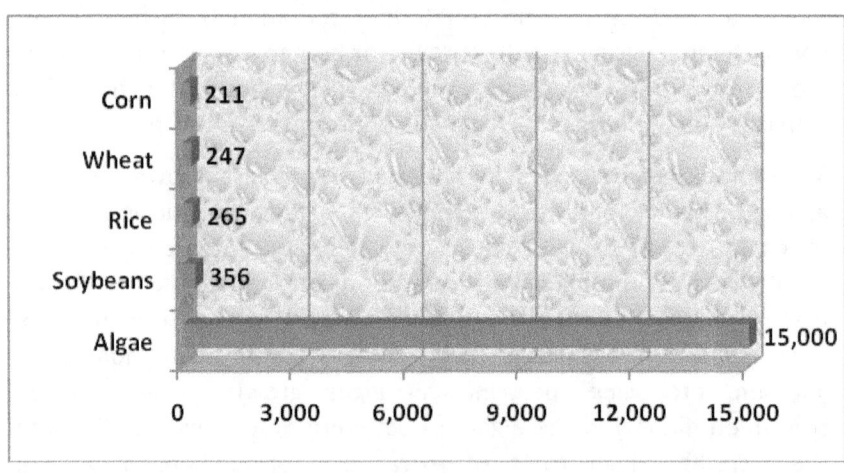

Currently, only certain species of algae are digestible by non-ruminants (animals such as cows with four stomachs) because the cell walls are too hard. Several research streams promise solutions to the cell wall issue. When production and cell wall challenges are solved, probably in the next three years, algae will be ready to serve as a sustainable food because it is robust, resource efficient and farmers can be trained to grow it easily.

Robust production

Green solar is incredibly robust because in closed loop algaculture systems that may look similar to an oversized aquarium, the plant grows independent of climate, weather or geography. Assuming variation in species, food and fuel can be produced all over the world, even in the north latitudes. Container-based production enables farmers to grow biomass with geographic independence – on mountains and deserts; lakes and the ocean. Farmers in unusual climates may select hardy species or use some form of protection such as greenhouses to enhance productivity. Canada recently announced three algal production projects using a combination of greenhouses and geothermal for heat.

Robust growth

Altitude. Algae grows everywhere on Earth including in snow on mountain tops, under the ice shelf in Antarctica and in hot deep ocean vents.

Latitude. Algae may double its biomass daily when there is sun so it has no growing season. Some algaculture systems may need heat and artificial light to produce effectively in winter.

Geography. Algae may be grown on mountains, deserts, rooftops, garbage dumps, wasteland or oceans.

Soil. Algae do not need soil for growth since it grows in waste, brine or salt water.

Temperature range. Various algal species grow in every temperature range on Earth. Producers may use greenhouses or other means to optimize temperature for maximum production.

Climate. Algae can be grown in a closed system independent of climate. In extreme climates, producers may need to grow a different species each season. Cold locations may find cold growing species or use geothermal heat to warm greenhouses.

Salt tolerant. Algae is not only salt tolerant but many species prefer saline growing mediums

Weather events. Closed systems are whether event independent. Storms or drought have minimal impact on production because as soon as the weather passes and the sun returns the algae begin to grow again.

Storms may slow or stop algae production during the bad weather but the growing system may begin again as soon as the sun returns. Extreme weather may kill the plants in an algaculture system. However, assuming some production culture was preserved in a sheltered place such as a box in the ground, farming could begin again. If weather extremes were expected, local growers may decide to stop growing biomass during the period or use shelters such as greenhouses to protect the plants.

Even a robust organism needs to be extremely resource efficient in order to be sustainable, especially in a global warming environment. Algae are resource efficient because they learned to survive in the brutally harsh conditions of early Earth.

Resource efficiency

Green solar produces with high resource efficiency.

Resource efficiency

Cropland. Algae need no cropland and prefer deserts where food crops do not grow. Marine species may be grown in estuaries, bays and oceans.

Space. Creates substantial biomass production in a very small space. Micro algaculture systems that support a family, cooperative or village may fit on a balcony, a roof top or other non-cropland site that offers solar exposure.

Air footprint. Creates a positive atmospheric footprint by sequestering CO_2 and water and giving off lots of O_2 as it makes high energy plant biomass.

Water footprint – creates a positive water footprint and cleans water from agricultural, industrial or residential waste streams. Creates a positive water footprint since the growing water can be recycled in closed containers. Some evaporation loss occurs in warm climates.

Soil footprint – Creates no soil ecological footprint since it is grown in closed containers and the fertilizer is recycled with the growing water.

In-situ fertilization – algae can use the waste streams from agriculture, especially dairy, as well as municipal and industrial wastes for fertilizer.

Inexpensive seeds – algae propagation using a variety of strategies is easy, fast and relatively cheap.

Fossil fuels. Demands minimal fossil fuel energy for biomass production. Solar or wind power may provide sufficient energy for production.

Fossil inputs. Consumes minimal fertilizers, herbicides and pesticides. No inputs are wasted since nutrients can be recycled with the growing water.

Transportation. Needs minimal transportation since it can be grown locally.

Algae's ability to use recycled fertilizers that exist in the place where they grow provides a tremendous advantage to solar farmers. Most farmers have available waste streams from human, animal and vegetative wastes on which algae can thrive. Rather than spending 25% of their production costs on fertilizer, algae may cut that cost to 5%. Algaculture typically needs some micronutrients and extra CO_2 for high production.

Algae propagation may occur by a wide variety of sexual and asexual methods. Since the organism multiplies so fast, a single solar farmer can create enough algal culture to supply 1,000 farmers with sufficient culture for start-up. Food crops require substantial human time and labor. Traditional farming requires a wide variety of human inputs and risks that are minimized by green solar.

Human efficiency

The tedious, heavy and risky work of traditional agriculture makes it impossible to produce food for many people with the most need. Green solar offer significant human efficiencies.

Human input efficiency
Heavy labor – requires few activities that demand physical strength.
Tedium – fast production with a light labor requirement avoids tedium. However, green solar production parallels a dairy in the sense that daily activities need to take place.
Continuous work – production may start and stop at the farmer's convenience.
Physical risk – creates minimal risk of accidents that could cripple or kill the farmer. Avoidance of agricultural chemicals minimizes health risks.
Psychological risk – creates less anxiety than traditional farming due to lack of dependence on rain or growing conditions.
Financial risk – minimizes farmers' financial risk because after the growing system is built, inputs are cheap at production reliable.
Labor hours – requires relatively few labor hours to produce a pound of protein.

Production period –requires days rather than an entire growing season, four months, to produce the first pound of protein.

Most green solar activities can be performed by either men or women because they are not physically demanding. Teenagers and older people can do much of the work since the labor requirement is on par with gardening. Basic production, harvest and sun drying involve relatively light work. More advanced uses such as pressing out fuel oil may necessitate more strenuous work but the timing of those activities can be scheduled.

Daily production, harvest and drying may take two hours a day, so it minimizes the tedium and avoids exhaustion. Production can stop and restart at the farmer's convenience with minimal productivity loss which represents a huge difference compared with food grains. Algaculture minimizes risk of physical injury because the equipment is simple and not dangerous and monotony and fatigue are not factors. Farmers do not have to handle poisonous chemicals. Daily production minimizes psychological and financial risk for farmers since there is low probability of crop loss. In the event of the loss of a culture, a new culture can restart the next day, with new production in a few weeks. After the capital cost of the growing system, farmers have minimal costs while producing high value food, fuel, fodder, fertilizers.

Diffusion

Even a highly productive plant must be really, really simple to grow because millions of farmers will need to adopt new methods. Jeffry Sachs proposes that an effective R&D model for the developing world is R3D; research, development, demonstration and diffusion.[212] R&D would make the plant available but adoption would require that the plant be very easy to demonstrate and to convey to others.

Everett Rogers studied the diffusion of innovation for decades and documented the diffusion of hybrid seed corn and the incredibly slow progress in diffusing new agricultural methods. He concluded that people's attitude toward a new technology is a key element in its diffusion. Roger's Innovation Decision Process states that innovation diffusion is a process that occurs through five stages: knowledge, persuasion, decision, implementation and confirmation.[213]

Jeffry Sachs' R3D model aligns with Rogers' findings which mean that simplicity will be critical for effective demonstration. A plant that grows sufficiently easy for wide diffusion must be so easy to grow that farmers and support persons such as family members can learning quickly. Of course, demonstration and diffusion can occur only after strong research and development have solved production challenges.

Really, really simple Constraints
The Learning Tools needed for Green Solar Diffusion

Learning curve – users experience a one or few day learning curve – they can see and immediately do with no-trial learning.

New stuff – users will have to learn only a few basic actions and a few new technologies.

Visual learning – users will be able to learn visually so language will not be an obstacle.

Simple tools and labor– users will be able to substitute labor and simple tools for technology.

Forgiving system – users will know how to make the plant grow and produce even if they do not understand how and why or how it produces.

A plant that is robust, resource efficient and really, really simple to grow presents a fascinating model for sustainable food production. Even though algae may be robust and resource efficient, adoption by farmers around the world depends on demonstration and diffusion. Models developed by E. F. Schumacher, *Small is Beautiful*, Jeffrey Sachs *Common Wealth: Economics for a Crowded Planet* and The Hunger Project provide templates for demonstration and diffusion.

Farmers may have to learn how to balance water acidity. This is equivalent to testing a swimming pool which is designed to kill algae with acid. Algal species are robust in many ways but are pH sensitive. Training materials will need to be available visually such as a simple picture book. Naturally, farmers will be able to ask questions about their concerns. Fortunately, farmers will be able to produce foods and possibly cooking fuel, without knowing exactly why the process works.

Farmers must be able to easily learn to grow the biomass and extract the oils for cooking stoves and the protein for food or fodder. Small green solar systems could be demonstrated a few hours. Farmers could learn to produce product in one day. With a smart rollout strategy, farmers could look at a demonstration unit and immediately begin producing algae. The roll-out model will also include Margaret Meade's concept of lateral learning where farmers help one another get started.

Chapter 9. Cultivating Abundance

The fall of fossil agriculture will occur in the lifetime of our current farmers' children because their parents will have extracted the last of the fossil resources that leverage industrial food production.

For thousands of years, coastal people have harvested seaweed for food, fodder, cooking fires, fertilizers and medicines. Indigenous people used seaweed for treatment of bruises, burns, cuts, skin irritations and indigestion. Dried seaweed has been used for centuries for trade because it is nutritious; easily transported long distances and has a long shelf life. When put in water, the original color, which may be purple, red, brown, yellow or green, reappears. Some seaweed is not eaten directly as food but is used to flavor salads, soups and stews or as a condiment for meats or rice.

Most the countries along the Pacific Rim have thousands of years of experience in using algae as food. Recently seaweeds have entered European, Canadian and California cuisine and are often regarded as an exotic component of the menu. Cook books are incorporating recipes using sea vegetables, Nori and other seaweeds grown organically in clean environments.

Some seaweed such as Nori, Kombu and Wakame offer high dietary fiber content. Most seaweed have a low fat content and are rich in vitamins such as niacin, vitamin B, but may lose their vitamins in drying and processing. Food researchers are searching for ways to retain these valuable vitamins and micronutrients because billions of people are deficient in vital vitamins. Raw seaweed contains appreciable amounts of essential trace elements such as manganese, copper, cobalt, iron, nickel and zinc. Consequently, many societies have used fresh seaweed as a fodder additive to keep their animals healthy.

On both coasts of North America, producers have begun cultivating seaweed in estuaries and bays and onshore in tanks called cultivated algal production systems, (CAPS). Seaweed and other algal foods markets are growing as people rediscover the good taste, texture and nutrition that were once a part of traditional diets. A recent Rotary meeting at a Kobe Steakhouse served three forms of algae at the luncheon: soup broth for taste and thickening, sea vegetable in the soup for texture and visual appeal and separately a cold seaweed salad for color, taste and texture. The Rotarians were surprised yet delighted with the colors, taste and texture provided by the seaweed.

Nori, probably the most popular seaweed, grows as a very thin, flat, reddish blade that grows naturally in most temperate intertidal zones around the world. Today Nori is farmed on the coasts of most continents because demand is far too high to be supported with natural stands. Nori is among the most nutritious seaweeds, with a protein content of 40-50% and is 75% is digestible. Sugars are low, 0.1%, and the vitamin content very high, with significant amounts of Vitamins A, C, niacin and folic acid. The shelf life of vitamin C can be short in the dried product. Most of the salt is washed away during processing to produce the sheets. The characteristic attractive taste of Nori is caused by the large amounts of three amino acids: alanine, glutamic acid and glycine.[214]

Nori serves as a luxury food and is often wrapped around the rice ball of sushi or rice with a slice of raw fish on the top. Toasting or baking brings out Nori's rich flavor and flakes are often sprinkled over boiled rice or noodles. Nori may be incorporated in soy sauce and boiled down to make the rich sauce. Nori is also used as an ingredient for jam and wine. Chinese cooks use Nori in soups and for seasoning fried foods.

Natural settings

Algae have been harvested in natural settings for centuries along coastlines and in estuaries and lakes. Various strains of the Spirulina plant are farmed in Africa, Lake Chad and in Mexico, Lake Texcoco. Villagers harvest about 100,000 tons of algae from the volcanic Twin Taung lakes in Myanmar where the biomass serves as a local food supplement, animal feed and is sold to distributors as a health food.

The abundance of algae globally would seem to make it a strong candidate for domesticated (farmed) production in natural settings such as ponds, lakes, estuaries, bays and oceans where it flourishes naturally. Natural production methods work amazingly well to support nature's food chain but are insufficiently reliable or productive for human food or energy production. In natural settings such as lakes, algae supports thousands of different organisms and cultivated production would deprive those creatures of food, and for many, such as pink flamingos, of coloring. Nearly all the natural stands globally have been over harvested so the only practical solution for producing sufficient human food is algal cultivation.

Nature's challenges

Science must overcome a non-trivial set of nature's challenges to produce food reliably and sustainably. Solar farms that produce domesticated algae for specific purposes such as food and energy must leverage nature's strengths while overcoming some significant limitations. The major constraint to culture stability is called the law of the minimum, which refers to nutrient limitation, Table 4.1. When any important nutrient for algal growth becomes unavailable in the culture, because prior algal cells have already consumed the nutrient, algae stop reproducing and growth in the culture stops. Core nutrients include light, CO_2, nitrogen, phosphorus, potassium, sulfur, and calcium as well as a variety of trace elements.

Algae grow so quickly that one or more nutrients are likely to be limited in the confines of a puddle, pool or pond. Since algae cannot migrate, the plants depend on nutrients coming to them and when those nutrients are deficient, the plant stops reproducing. SAFE production overcomes nutrient limits by assuring the culture provides a sufficient supply of critical nutrients.

Table 9.1 Algal Growth Limitations in Natural Settings

Challenge	Natural setting limitations
Law of the minimum	Algae growths quickly as long as the proper set of nutrients are available locally. When one nutrient becomes deficient, the entire algal culture stops growing.
Extremely limited light interchange	Sunshine or artificial light, the critical source of energy for algae production can be captured only from directly above in most cultures which significantly limit the number of photons that can be absorbed by algae.
Shading	Algae grow quickly in all directions but the cells that grow above prior cells tend to shade the older cells, which causes them to stop growing.
Contamination by other microorganisms	Other microorganisms that either compete with algae for nutrients or consume algae directly can significantly reduce culture growth and productivity.
Invasion by weed algae	Weed algae can enter a culture and propagate quickly to dominate the culture. Weed algae may substantially diminish the biomass value because it changes the composition of the algal biomass.

Algae typically absorb sunlight only from the tiny (few inch) interface between the atmosphere and surface of the water, which substantially limits its photosynthetic efficiency. A tree, for example, absorbs photons from hundreds of square feet of leaves that tower high above the ground. In natural settings, algae lack the vertical dimension which limits photon absorption. Photosynthetic efficiency can be increased by building vertical CAPS made of transparent materials that significantly increase solar exposure.

Shading creates another extreme limitation to growth in natural settings. As new plant cells grow, they tend to shade prior plants, which first reduces, then eliminates the solar energy necessary for photosynthesis. In natural settings, it is not unusual for algae to grow successfully only in the top two inches of the water where they can absorb sunlight. Shaded cells typically stop growing and may fall to the bottom of the water column. Others tend to drift in the water and are attacked by bacteria which consume the algae and create the unpleasant smell associated with stinky ponds. SAFE production minimizes or eliminates shading by continually mixing the algal culture. The same effect can be accomplished by growing algae in slowly running water or in the oceans where currents create replenish nutrients such as in an upwelling.

Contamination from other microorganisms that feed on algae such as bacteria and rotifers enter natural algal stands and can propagate quickly and consume huge amounts of algae. These microorganisms may already be in the water, be blown in on the wind or be carried by insects or birds. Rotifers, like algae, can propagate exponentially and may consume 90% of an algal stand in one morning.

Weed algae present a similar problem. In nature, weed algae enter a pond and if conditions are right, may grow to dominate the culture. This presents no problem in nature because a diversity of algae species simply attracts a wide variety of hungry feeders. However, in domesticated solar gardens, weed algae are likely to change the biomass composition, similar to weeds in a traditional garden. A change from high protein to high carbohydrates may make the biomass less valuable to the community. SAFE production needs to monitor and manage weed algae invasions which is accomplished by maintaining culture parameters such as temperature and pH.

Nature's limitations create a difficult but manageable set of challenges to domesticated algae production. Anyone serious about algae production should read the excellent source published by the Phycological Society of America, *Algal culturing techniques* edited by Robert Andersen.[215]

Seaweed – mariculture

Various species require different farming methods. Some seaweed can be grown with one-step farming through vegetative propagation (cuttings) while most need multistep cultivation. *Eucheuma, Kappaphycus, Chondrus,* and *Gracilaria* are propagated vegetatively. *Porphyra, Enteromorpha, Laminaria, Undaria* and others are started from spores.[216] Seaweed typically shed spores which settle on mollusk shells or rocks. When cultivated, spores are carefully placed on shells and allowed to grow in a controlled environment until they are strong enough to be placed in the open environment.

Farming seaweed generally occurs in shallow bays or estuaries where ocean nutrients naturally feed the plants. Some species do well in cultivated tanks and ponds such as *Chondrus crispus,* which is cultivated in tanks in Canada and in Portugal. *Gracilaria* is grown in tanks and raceways in Israel and in human-made ponds in Taiwan and Thailand. Over 100 seaweed species have been tested for farm cultivation but only a dozen are being commercially cultivated. Most seaweed cultivation occurs in the eastern Pacific Rim countries.

Porphyra, known as Nori, is the most valuable farmed seaweed in the world and includes at least 33 species. *Porphyra* reproduces by both sexual and asexual modes of reproduction and farmers typically use spores for propagation. A four step cultivation process which includes culture of conchocelis, seeding of culture nets with conchospores, nursery rearing of sporelings and harvesting.

Mariculture typically involves a set of steps similar to nursery production for land plants. First, the farmer induces the seaweed to propagate. Then the spores or seedlings mature in a controlled environment and are then planted on a shell, stake or line where they grow to a level of maturity that they can survive in the wild. The immature plants are placed in position where they have access to sunlight and nutrients. Seaweeds are grown on bamboo or mangrove sticks, monofilament fish line or ropes made of hemp, plastic or materials. Mariculture in ponds or tanks follows the same protocol but farmers have considerably more control over production variables and doing this which is critical for products that enter the food system.

Several countries including China and Taiwan have developed polyculture production where several species of marine organisms are grown in the same pond as the seaweed. Shrimp, clams, crab, fin fish and prawns are grown in various combinations or even all together. The fish and crustaceans provide extra income to seaweed farmers and an efficient usage of the ponds.

Mariculture shares similarities with land-based farming as producers must continually weed out undesirable seaweeds, maintain the growing support lattices, replace lost seedlings and remove competing organisms called grazers. See urchins, starfish and finfish need to be removed because they can consume large quantities of seaweed biomass.

Mariculture typically produces one to three crops a year depending on the species. Growth from adolescence to maturity typically takes 2 to 3 months. The growing season usually aligns with land plants where maximum solar energy is available and water temperatures are optimum for growth. Many seaweed species such as kelp are distributed widely along the world's coastlines but cold water kelp is a different species than tropical kelp. Both species may share many attributes including biomass composition.

Cultivation in ponds

Pond systems in the U.S. were first developed for water treatment. The recovered biomass was converted to methane and burned as a local source of energy.[217] When energy was cheap, the energy value of algae was considered incidental.

Most microalgal production today occurs in open ponds, called "slime ranching," because ponds are cheap to build and operate. However, open pond cultures are limited to a relatively small number of algal species and ponds make it difficult to control invasion from weed algae and predators such as bacteria, rotifers, viruses, fungi, and zooplankton.[218] Open ponds are vulnerable to contamination from dust, windborne organisms, insects and birds. In addition, ponds are limited geographically to tropical and subtropical areas with warm temperatures, low rainfall and little cloud cover.

Evaporation represents a critical limitation to open ponds, especially in hot climates. An open pond may lose half its water in a growing season. Evaporation increases retained salts which may affect the stability of the culture. Some open ponds use seawater, waste or brine water which often makes the water free but does not slow evaporation.

Most microalgae need light and carbon dioxide but they vary substantially by specie in nutrient and environmental requirements. Some species grow well in unlined ponds in Australia but the same species may not flourish in unlined ponds in India or China. Consequently, local conditions often dictate the design and construction of open ponds as well as species selection and production. Relatively little published research is available on open pond production because production methods and productivity are protected as trade secrets.

Algal ponds typically are shallow, one to three feet deep, in order to maximize cell access to light. Algae grow quickly and new cells shade older cells so in unmixed ponds will have growth only in the top two inches of the water column. The ponds are typically mixed with a paddle wheel that keeps the culture moving around the raceway or in a circle. Water movement around the raceway creates enough turbulence to bounce cells to the surface where they can absorb photons. Large ponds used for municipal water remediation bubble a mixture of CO_2 and air to move the water.

Large ponds tend to be less expensive per unit area to construct and operate. Size affects water circulation, operating costs, mixing systems and species selection. Mixing gives cells access to light, prevents cells from settling to the bottom and avoids thermal and oxygen stratification in the pond. Effective mixing systems can create high cell densities which reduces harvesting cost. Harvesting typically occurs with centrifuge, filters or flocculation.

Commercial algal production in outdoor ponds is vulnerable to weed algae invasion from competing local species. Producers grow *Spirulina* at high bicarbonate concentrations which yield a high pH that discourages competing species. *D. salina* is grown in a high saline culture that minimizes contamination by unwanted algal species. Outdoor ponds often use a batch culture approach to control contamination where the culture is restarted at regular intervals with a fresh single-specie culture.

Closed systems

The term "closed system" is a misnomer because algae predators and weed species invade any outdoor growing system. It may be better to think of a cultivated algal production system, CAPS, as an arrangement that gives the farmer more control over production parameters and more but not perfect control over contamination from invasive algae and predators. Farmers typically control predators and weed species with a combination of parameters including pH, temperature and nutrients.

The commonly used term in the algal industry for growing systems, photobioreactor, is avoided here because people will not want to eat food produced in a "reactor." In addition, the term bioreactor has become synonymous with wastewater treatment facilities.

Acrylic that is ultraviolet (UV)-stabilized is typically used for CAPS construction because compared with glass it is cheaper, stronger, lighter, more flexible and easier to work with. Assembly of the biofactory requires the integration of the various mixing, monitoring and controlling subsystems.

Fermenters

Most algae species are autotrophs and are able to use solar energy directly to produce organic substrates that store chemical energy from inorganic CO_2. Many species can also function as heterotrophs where they are able to metabolize organic substances to create and store the chemical energy needed for the lifecycle. Heterotrophic algae can be grown in large containers called fermenters without light and are typically fed sugar as their primary energy source.

Fermenters offer a number of advantages including that considerable published information is available as well as tested commercial growing systems. Heterotrophic growing systems typically have lower operational costs than light-based systems.[219] Algal biomass production without light usually uses a pure algal strain called an axenic culture that is free of other contaminating organisms. Molecular biologists have recently genetically transformed autotrophic algae that feed on light energy to heterotrophs. These transgenic cells thrive on sugar in the absence of light. The dual algal growth mode enables more flexibility in designing and operating growing systems.

Production findings

Decades of algal research and field production experience have yielded the following general insights on algal production.

Table 9.2 Primary Production Variables

Variable	Insight
Species variation	Each species reacts differently to variations in growing parameters and each strain within a species may have different reactions to changes in the culture.
Temperature	The optimal temperature range for growing algae is 55° to 95° F, although some species can grow at almost any temperature above zero. Most species

	have a maximum productivity sweet spot and slow growth outside their favored temperature range.
Light	Production is optimized with intermittent light and dark cycles of 1 to 15 seconds. Container shape, mixing method and speed as well as light sources determine light cycles.
Stress	Stressing algae by withholding nutrients or changing other growing parameters causes the plant to implement a defense strategy for survival and to change the biomass composition.
Fouling	Some species stick to the sides of growing containers, blocking light to other cells while others do not. Sticky algae make poor candidates for cultivation.
Salt	Algae are vigorous in cultures with a wide range of dissolved salts and waste, including heavy metals. Algae often grow most productively in water slightly more saline than their natural habitat
pH – acidity	Many species have a relatively narrow acceptable acidity range measured by pH. Acidity may be used to control opportunistic algal contamination.
Shear stress	Algae are relatively delicate soft cells and suffer from shear stress if the culture is mixed too rapidly. Shear stress also diminishes culture productivity in vertical water columns above about 4 feet. Fortunately, engineers have found a number of ways to minimize shear stress such as creating a zigzag similar to the flow of water in a radiator in vertical systems.

Light

While light is the most important parameter in system design, it is difficult to measure because as the cell concentration changes, the light requirements change. Most species are relatively light-sensitive and react negatively to too much light, which causes photoinhibition and slows growth rates. Too little light or not enough light at the right time also diminishes productivity. Light may be supplied continually or in light dark cycles.

CAPS design needs to maximize light penetration and minimize reflection and refraction. For this reason, curved surfaces like tubes have less available light than flat surfaces. Some designs use mirrors, refractor's, parabolic light collection devices, fiber optics and light guides. The intensity of light trading the culture drops quickly because the chlorophyll in the algal cells absorbs the light. With high cell concentrations, most of the light will be absorbed in the first two inches leaving remaining cells in darkness. Mixing and circulation systems are designed to balance access to light. Therefore, CAPS thin containers, about six inches, out-produce thicker containers.

CAPS have a photic zone close to the illuminated surface and a dark zone away from the light. Algae absorb photons in the light sequence and then need time in darkness to metabolize the energy. The light dark sequence may be anywhere from 4 seconds to 100th of a millisecond. Researchers have worked out optimal light dark cycles for many types of algae. Two parameters are important for light dark cycles; the light fraction, the ratio between the light period and the cycle time and the length of the light dark cycle. The biomass yield and growth rate is affected primarily by the light fraction while cycle duration has minimal impact.

Most cultures need added CO_2 because even bubbling air through the culture limits available CO_2 and slows growth. Common practice is to combine air and CO_2 as part of the mixing system. Algae produce so much O_2 that the oxygen needs to be removed from the culture. Producers typically use oxygen sensors and purge the culture of oxygen occasionally. Similar sensors are used to control temperature and pH.

Mixing

In natural settings, algae grow so fast that the new growth occurring closest to the sunlight tends to shade prior cells. Cells that cannot get light stop growing, stop producing biomass and stop propagating.

Adding to the complexity of shape, different algal species exhibit variation in their appetites for light and dark. In general, algae seem to grow most productively in a setting where they have access to light about every 10 seconds. During each brief dark period, algae seem to be digesting the light. Thin tanks tend to out-produce thick tanks because algal cells get more solar exposure. Some species are sensitive to the harsh light and grow best in soft light.

Theoretically, faster mixing in larger tanks should enable thicker tanks and high production. However, mixing above a relatively modest velocity threshold stresses and injures the soft plant cells, shear stress, and may create instability and the plants may stop propagating. A system that receives only intermittent mixing produces biomass but at a slower rate. Therefore, constant, relatively gentle mixing aids to productivity. In closed systems, compressed air and CO_2 is bubbled through the tank and the bubbles provide turbulence for mixing.

Water types

Algae can grow in salt, brackish, human waste or industrial wastewater. Land plants die in saline water but algae have no problem with salt or heavy metals because they have no roots. Some species simply absorb the salt or heavy metals from the water leaving it clean enough for crop irrigation. After the algae are harvested, the salt, heavy metals or other pollutants may be recovered as coproducts.

Protein recovered from wastewater production systems may meet food quality standards when solar heating or other protocol kills contaminants. However, these systems may not meet consumer perceptions of cleanliness. Organic fertilizers used on terrestrial food crops are made from the components of wastewater – primarily urea. The total growing process for food requires much higher standards of cleanliness than production systems optimized for oils or other coproducts. Production standards rise further if the producer grows food with an organic label. No organic standards have been proposed yet for water-based food production.

Nutrients and CO_2

Algae consume about the same nutrients in the form of fertilizers as land plants – only in smaller proportions. Nutrient assimilation is often affected by the presence or absence of other factors in the culture including culture temperature, available light, pH, sodium, potassium and magnesium. Since the biomass grows so fast to very high densities, the plants need nutrients regularly. Corn for example, assimilates only about half of the nitrogen fertilizer put on a field. Algae may not absorb all the available nitrogen in a growth culture but the remainder may be recovered and recycled, along with the water for the next growing cycle.

About 50% of algal biomass consists of carbon so a sufficient supply of carbon is vital for successful cultivation. Carbon can be supplied as an inorganic substrate in the form of gaseous CO_2 or in the form of bicarbonate. Organic carbon sources are mainly sugars or acetate. Algae can absorb CO_2 from the water but most water carries little CO_2, about 0.5% in solution and 0.03% in air – both densities are too low to sustain optimal growth and productivity. Growers must also remember that CO_2 diffuses 10,000 times faster in air than in water. Most growing systems use supplemental CO_2 that is either bubbled into the tanks as a gas or added to the tank has a solid, typically a carbonate. Siting biofactories near CO_2 sources such as power plants or beer manufacturing plants saves input costs and sequesters CO_2.

Besides CO_2, nitrogen is the most important element in algal growth since more than 10% of the biomass consists of nitrogen. Nitrates, ammonia and urea are most commonly used as a nitrogen source. Other major elements include phosphate, potassium, magnesium, calcium, sodium and sulfate.

Phosphorus is a major nutrient and is essential for almost all cellular processes because it works on biosynthesis of nucleic acids and energy transfer. Like phosphorus, sulfur is vital to all cells because it serves as a constituent for several essential amino acids (methionine, cysteine and cystine) and vitamins. Calcium ions also play a role in cell walls and form the skeletons of certain algae. Magnesium is critical to all photosynthetic organisms because it plays a central role in the operation of the chlorophyll molecule. Iron is critical for growth and plays an important role in algal metabolism and nitrogen assimilation. Similar to land plants, bleaching or yellow coloring in algal cultures typically indicate iron deficiency in the growing medium.

Trace elements provide critical micronutrients that are required in micro-, nano- or even picograms per liter. Trace elements influence growth in a representative number of species and show a direct positive physiological effect on algal growth. Trace elements generally cannot be replaced by another element and the culture may show reversible signs of deficiency, e.g. color or composition, in cultures lacking the element. The major trace elements in algal cultures are manganese, nickel, zinc, boron, vanadium, cobalt, copper and molybdenum. Some trace elements have a narrow range between offering important nutrition and cell toxicity. For example, copper is essential at extremely low concentrations but acts as an algal poison at high concentrations. Copper is often used in controlling algal blooms.

Algae are able to concentrate certain heavy metals in their cells at more than 1000 times ambient levels. This ability enables algae to detoxify political industrial wastewater with heavy metals. Algae also can remove pharmaceuticals from municipal water. Once chemicals are concentrated in algae, they may be removed in processing. Technically, it is easier to separate chemicals in a solid, concentrated algal biomass, than when the chemicals are widely diffused in water.

Substantial variation in growth medium recipes occurs for various strains and species and may vary based on local growing conditions. Producers also sometimes stress algae at a certain time in their growth cycle by adding or subtracting a specific nutrient. When algae are stressed by limiting a nutrient, their defense mechanism activates and they begin storing the energy they need to survive. Defense strategies include faster or slower growth, storing more lipids, proteins or carbohydrates or creating novel compounds.

Algal cells use elements in many ways as illustrated in Table 8.7. [220]

Table 9.3 Use of Elements in Plants

Element	Function in plants
	Non-mineral nutrients – source: air and water. Plants use solar energy to change CO_2 and H_2O into chemical energy stored in starches and sugars.
Oxygen, O	Supports photosynthesis, respiration and plant growth.
Hydrogen, H	Supplies the H for the creation of hydrocarbons.
Carbon, C	Supplies the C for the creation of hydrocarbons.
	Macronutrients
Nitrogen, N	Nitrogen plays key roles in all living cells and is a necessary part of all proteins, enzymes and metabolic processes involved in the synthesis and transfer of energy. • Nitrogen is a part of chlorophyll, the green pigment of the plant that is responsible for photosynthesis. • Helps plants with rapid growth, increasing seed and fruit production and improving the quality of leaf and forage crops.
Phosphorus, P	Like nitrogen, phosphorus is an essential part of photosynthesis.

	• Involved in the formation of all oils, sugars, starches and cellular growth.
	• Helps with the transformation of solar energy into chemical energy, plant growth and maturation and helps plant withstand stress.
	• Encourages rapid growth, blooming, fruiting and root growth.
	• Drives ATP, DNA and phospholipids.
Potassium, K	Potassium is absorbed by plants in larger amounts than any other mineral element except nitrogen and, in some cases, calcium. • Helps in building plant proteins. • Essential for photosynthesis, fruit quality and protection from disease.
Magnesium, Mg	Magnesium is part of the chlorophyll in all green plants and essential for photosynthesis. • Helps activate many plant enzymes needed for growth. • Oxygen evolving complex of photosystem II • Aids in cell wall formation
Sulfur, S	• Essential plant food for production of plant proteins. • Promotes activity and development of enzymes and vitamins. • Helps in chlorophyll formation. • Improves root growth and seed production. • Helps with vigorous plant growth and resistance to cold.

	Micronutrients
Boron, B	• Helps in the use of nutrients and regulates other nutrients. • Aids production of sugar and carbohydrates. • Essential for seed and fruit development. • Sources of boron are organic matter and borax
Copper, Cu	• Important for reproductive growth. • Aids in root metabolism and helps in the utilization of proteins. • Plastocyanin, cytochrome oxidase
Chlorine, Cl	• Aids plant metabolism. • Oxygen production in photosynthesis.
Iron, Fe	• Essential for the formation of chlorophyll.
Silicon, Si	• Nitrate reductase • Supports cell wall formation.
Sodium, Na	Chlorophyll functions
Magnesium, Mg	• Operates with enzyme systems involved in the breakdown of carbohydrates and nitrogen metabolism.
Molybdenum, Mo	• Helps in the use of nitrogen.
Zinc, Zn	• Essential for the transformation of carbohydrates. • Regulates consumption of sugars. • Part of the enzyme systems which regulate

	plant growth. • Critical for plant reproduction.
Cobalt, Co	• Vitamin B_{12}
Vanadium, V	• Bromoperoxidase and some nitrogenases
Bismuth, Br	• Halogenated compounds
Iodine, I	• Protects plant with antimicrobial, anti-herbavore or allelopathic functions

Some elements are provided to algal cultures in trace amounts because very little is needed by the cells.

Delivery and timing

Nutrients are typically delivered into the growing medium either constantly or intermittently during the day when algae are growing. There is no need for nutrient delivery at night when algae are resting. Some species like burst feeding where large amounts of nutrients are introduced at one time. More commonly, most algae prefer drip feeding where nutrients are presented in a fairly constant pattern throughout the day.

Algae display variations in the speed of growth at various times during the day. A common pattern is a steady increase in the speed of growth throughout the morning until about midday. The plants do not exactly take a siesta but they tend to slow their growth as the afternoon progresses. Growth tends to be slower in the early morning and late afternoon when the plants have less access to sunlight.

Harvest and demoisturing

Considerable variation in harvesting reflects variations in species, size and available technology, Figure 9.1. Harvest in natural settings often use floating or settling. Algae that float often clump can be skimmed off the surface. Settling allows the culture to sit quietly overnight and the cells simply sink to the bottom where they can be removed. Harvested algae may be poured into a basin formed in sand, then covered with fabric allowing the water to flow through to the sand leaving the algae to dry. The biomass can be scraped off the fabric.

Figure 9.1 Harvest, Separate and Dry

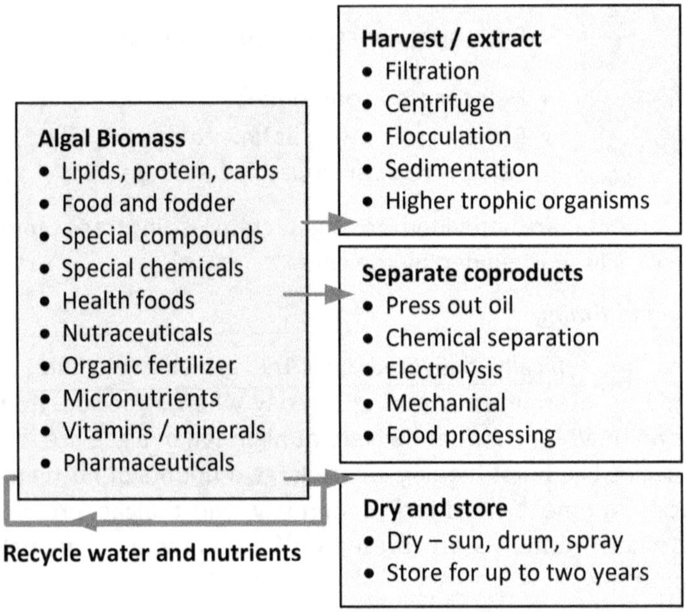

Harvest / extract
- Filtration
- Centrifuge
- Flocculation
- Sedimentation
- Higher trophic organisms

Algal Biomass
- Lipids, protein, carbs
- Food and fodder
- Special compounds
- Special chemicals
- Health foods
- Nutraceuticals
- Organic fertilizer
- Micronutrients
- Vitamins / minerals
- Pharmaceuticals

Separate coproducts
- Press out oil
- Chemical separation
- Electrolysis
- Mechanical
- Food processing

Dry and store
- Dry – sun, drum, spray
- Store for up to two years

Recycle water and nutrients

Filtration may use fabric, cheese cloth or microscreen filters. In high productive systems, harvesting occurs once a day at maximum cell density which is typically late morning. A third to one half of the algal cells may be removed from the growing medium.

Some species are so small they require flocculation. Flocculation is derived from the word floc or flakes of material. When a solution is flocculated, the small solids are formed into clumps of aggregate which are easier to see and to remove with filters or screens. Flocculants are commonly used in water treatment to improve sedimentation and filterability of small particles. Alum, ferric chloride and chitosin are common flocculants. Flocculation may be too expensive for small-scale algal production. Froth flotation, another harvest method, aerates the water into froth and the algae are skimmed off the froth. Interrupting algae's CO_2 supply can cause algae to flocculate on its own, which is called autoflocculation.

Some algae secrete various products, especially oil. A production system using secreted oil can harvest the oil by skimming the top of the culture. This method does not require harvesting or processing the algal biomass. Reports on secretion show some advantages but these systems do not produce as much oil or algal coproducts, including protein and micronutrients.

After harvest, algae are scraped off the filter. In some cases, algae are dried in the sun similar to grapes that are dried to make raisins. Other production systems use a centrifuge to demoisture the algae down to about 5% water. The extraction plan may be driven by convenience or available technology.

Trophic harvest

The simplest harvest method avoids harvesting algae directly by diverting part of the algal culture to a container with filter-feeders such as bivalves that feed on the algae. These larger trophic level creatures, higher on the food chain, have a natural affinity for algae and relish the work of harvesting. Brine shrimp work well for trophic harvest and are easy to harvest with a net. Brine shrimp can be squeezed to extract oil and the remaining protein biomass eaten or fed to animals. About 30% of current algal production targets fin and shellfish in aquaculture systems such as hatcheries and fish farms.

An aquaculture producer may begin a batch culture by filling a disinfected tank with pasteurized, filtered water and inoculating it with a 1:20 or 1:60 ratio of algal cells to water. The producer may keep a dozen strains of algae with different composition profiles of lipids, protein and carbohydrate in order to provide a balanced diet to the fish. These strains are available for purchase from several sources if the producer loses a strain due to contamination or system crash.

The algal batch culture stays in a controlled environment with light and warmth and an optimal nutrient mix. Algal cells multiply rapidly and in about ten days the batch changes from a light tint to a dark cloud that light cannot penetrate. The algal culture has matured and the culture is fed to fish. The containers are cleaned and the process repeats.

Some producers use a continuous algal production system where the algae are cultured in a greenhouse. The process begins with inoculation similar to the batch method and the fish are raised in carefully monitored containers with a series of input and output pipes continuously bringing in water and their algal food. The flow of water, CO_2 and nutrients in separate containers maximize the growth of the algal culture and about 25% of the culture is directed to the fish tank each day. Species selection and algal density allows the delivery of high fat or high protein cultures based on whether the fish are at the larvae, fry or juvenile stage of growth.

Trophic harvest simplifies one of the more difficult challenges to algal production, extraction, because the larger creatures can be easily harvested. Trophic harvest also eliminates component extraction if the fish are high enough on the food chain to be eaten directly.

Component extraction

Ideally, component separation enables the maximum number of components to be extracted from the biomass. Algal oil extraction methods vary from hand or mechanical pressing to chemicals and even ultrasonic sound that bursts the cell walls releasing oil.

Removing algae components from the algal biomass presents an interesting set of possible alternatives. Algae vary widely in their physical attributes so various press configurations such as screw, expeller or piston are matched for the type of algae. Mechanical crushing is often used in combination with chemicals. Usually, the highest value product is extracted first. Standard food processing methods typically work acceptably for the target components.

Quality control

Measurements of process quality vary with the goal for the system. Optimizing food production requires substantial monitoring, testing and assurance that the process meets FDA and sometimes organic food standards. Automated production systems enable quality control checks continuously for all of the critical variables. Quality control may include monitoring for:

- **Biomass** density, color, size, structure and vitality
- **Water** temperature, Ph (acidity), dissolved O_2 and CO_2
- **Water** quality and dissolved salts and possibly metals
- **Mixing** velocity and turbidity
- **Nutrient** availability for all important nutrients
- **Contamination** from weed algae or predator inasion

Measurement of various component parameters needs to occur through the growing process as well as harvest, oil extraction and component separation.

Nutrient sources

The cost of algal production will be minimized if growing systems can be sited near sources of nutrients. Nutrient input cost is not large but getting most of nutrients free substantially reduces operational cost. Major sources of CO_2 include power and industrial plants, waste treatment facilities, cement plants, breweries and heavy users of natural gas such as restaurants.

New methods of nutrient capture will benefit solar farmers. For example, cars produce about a pound of CO_2 per mile. If someone invents a carbon capture filter for vehicle exhaust pipes, there will be a nearly limitless supply of low-cost CO_2 for growing algae. The filter would have to be easily removable because each 100 miles traveled would add 100 pounds to the vehicle's weight and reduce gas mileage. Green consumers may endure the extra hassle if the process were safe, easy and fast. The CO_2 filters could be recycled at service stations and used to feed green solar production systems.

New methods of creating low cost plastic for algal culture containers will also be a significant breakthrough. An ideal outcome would be for solar farms to create a product line that included algal plastics so that the containers in which the biomass grows were produced by algae. An integrated algal production system might employ a few solar farms to mass produce nutrients and plastic that would serve hundreds of green solar gardeners. Similar mass production of component parts, microscopes, monitors, mixers and other items necessary for algal production will be necessary to bring construction costs to practical levels.

Production size

Most of the planned production for algal biomass in 2009 targets large green solar farms of several hundred acres to produce algal oils that will be used as liquid transportation fuels. Algae farmed for energy promise extremely high return on investment and represent the only algal application currently receiving public or private financing. Of the several Algal Biomass Summits held in North America, there has not been a single presentation on algae as a food source. However, large-scale algal oil production will provide thousands of tons of protein cake left over as a coproduct. Also, the substantial R&D that is not held proprietary will benefit green solar farmers who want to grow food.

The algal industry will probably develop similar to traditional farming with small, medium, large and mega-farms. Large firms will focus on the production of liquid transportation fuels while developing markets for a diverse array of potential coproducts. Large firms will require huge investments of hundreds of millions of dollars and will concentrate production technology with a relatively few suppliers.

Small and medium sized solar farms or gardens offer a distributed production model that has many advantages described in *Green Solar Gardens: Algae's Promise to End Hunger*. The knowledge and capability for green solar gardens can be distributed globally in a manner that will allow people to produce food, food ingredients, micronutrients, fodder for fish, fowl and animals, algal oil for cooking and heating fires and rich organic fertilizer locally. Green solar gardens can also be used to clean brine or wastewater.

Both large and small algal production systems will reclaim organic nutrients from human and animal waste streams and reuse them in SAFE production.

Chapter 10. The Rise of Abundance

*He that will not apply new remedies must expect new evils;
for time is the greatest innovator.*

– Francis Bacon

Our world desperately needs to apply a new approach to food production that does not rely on fossil resources because time is not on our side. Abundant agriculture with SAFE production offers solutions to many of the difficult challenges facing industrial agriculture. The rate of fossil resource consumption combined with perilously low reserves and the net zero export problem should make the case for urgent action to develop a food and energy supply built on a foundation of plentiful and cheap resources. In spite of the obvious logic, there are strong headwinds on the path to abundance that can be resolved with creative solutions.

Inertia

Humans have been moving along the land-based agriculture path for more than 11,000 years and will continue growing food crops in soil. Traditional agriculture takes its name from the habitual process of land-based crop cultivation and ingrained habits built on thousands of years of experience are very hard to break. Industrial agriculture enjoys intensive support worldwide that is unlikely to change in spite of obvious facts. Less than 1% the U.S. population practices farming but roughly 25% are employed in agribusiness, the management of food, fiber and natural resources from field to fork.[221]

Abundant agriculture needs to be positioned in a way that augments and benefits industrial agriculture while providing no resistance. Fortunately, green solar offers a number of benefits to traditional farmers.

Modern farmers face huge costs in disposing of their waste streams, especially in livestock and dairy production. Farmers need cheap freshwater and green solar production systems can clean waste or brine water sufficiently for irrigation. Farmers need affordable rich organic fertilizer which can be produced as a coproduct of water remediation.

Imagine a scenario where livestock farmers co-locate a green solar production system on their farm that reuses nutrients from the waste stream, yields clean water and organic fertilizer for their fields. Some farmers may grow algae to produce diesel fuel for their tractors and trucks. Several innovators, Including Ben Cloud at XL Renewables, are working on CAPS designed to address the multiple needs of modern farmers. Once the economics of green solar production are demonstrated, modern farmers are likely to become the dominant early adopters because they will receive so many benefits.

Modern farmers are ideal green solar producers because they bring a lifetime of experience in the husbandry of plants. Farmers are famously inventive and each will tweak the production system to maximize production in the local setting. Not only will farmers continually upgrade green solar production systems but they will find new coproducts that benefit their farm, family and community.

Agricultural policy

Farm policies in the U.S. and globally are entirely focused on industrial agriculture with farms growing land-based crops. All the subsidies in place to support crops, water and energy center on traditional crops. Agricultural lobbies have extraordinary influence in legislation that not only to support farmers with production benefits but exempt farmers from their pollution obligations such as the Clean Water Act and the Clean Air Act.

What America needs is an agricultural secretary not from the Corn Belt who has courage to make a very simple policy statement:

> Food and energy subsidies should benefit the production of sustainable and ecologically positive production.

Subsidizing ecological destructive food and energy production that will deprive our children of fossil inputs is antithetical to American values.

Consider that algal oil production on desert land in one-half an Arizona county could produce more energy than the entire ethanol production in 2009. Switching from corn to green solar production would save 40 million cropland acres, two trillion gallons of fresh water, six billion gallons of fossil fuel and significant pollution.

Failing an agriculture secretary with sufficient courage, a number of other strategies could change U.S. farm policy.

- High school and college students might lead the debate regarding the value they see in their fossil resources that are currently being wasted. Content for those debates are available at a number of food sites including www.greenindependence.org where teaching materials for the entire *Green Algae Strategy* series including *Crash!* are available in color PDF for free download for students, teachers and food and energy policymakers.
- Fishermen in Louisiana and Mississippi might sue the EPA, DOE, USDA and farmers for failure to perform their duties in regulating agricultural pollution. Fishermen in the Chesapeake Bay and other estuaries that are too polluted to support aquatic life may also join the action to reduce pollution.
- Texas cattlemen have a cause of action which was articulated by Governor Perry in his appeal to the EPA to reduce ethanol mandates because increasing feed prices due to ethanol production, eflation, were driving cattlemen out of business. Of course, the EPA ruled in its own favor and denied Governor Perry's request just as the EPA ruled against California when it sued the EPA for clean air on behalf for their food producers.

- Mississippi farmers and politicians are considering suing upstream states and federal agencies over the severe water and soil pollution that has decimated crop production in many areas caused by discharge of agricultural chemicals.
- A national referendum on ethanol production would end subsidies and import tariffs because most Americans know that ethanol production makes no scientific or practical sense.

In spite of public consensus, ethanol production targets and subsidies are increasing rather than decreasing. Ethanol refining subsidies, intended to create many new businesses and jobs, have quietly been siphoned off by one company that owned 70% of ethanol refining capacity, Archer Daniels Midland. AMD's CEO Dwayne Andreas "paid to play" by giving millions to politicians in both parties in the 1990's. Those millions have returned billions to ADM as subsidies have continued to flow to ADM for over a decade.[222] The *New York Times* reported that ADM will receive $1.3 B from ethanol subsidies in 2007 and stands to receive more in 2008 and each year going forward.[223]

Thomas Jefferson warned that the concentration of power in the executive branch could lead to corruption unless there was full and vigorous scrutiny of appointments. Appointments would be auctioned to the bidder with the highest business interest. "Withdrawn from the eyes of the people," Jefferson reasoned, "the appointments may be secretly bought and sold, as at a market."[224] As Jefferson predicted, many of the most important EPA, DOE, FDA and USDA positions have been filled with lawyers and lobbyists representing the worst polluters in their industries such as ADM and Big Oil.

Make the intangible tangible

The most critical hurdle to public support for change in food and energy policy lies in the fact that fossil resources are currently intangible. When important issues are out of sight, they tend to be out of mind as well and that makes it impossible to create a feeling of urgency. The Dalai Lama' warning about the hidden threat from soil loss is as apt for water and inorganic chemicals as it is for soils – when we discover they have crashed, it will be to be too late.

Most people are unaware of how perilously close we are to running out of fresh water for irrigation, fertile soils or mined agricultural chemicals because these resources are not visible. These critical resource issues are not just below most people's radar, they do not even exist as a concern because they have not been discussed outside of scientific circles. Strong metrics can supply the visual evidence in graphics and pictures that will share the story and prompt broad action. Fortunately, these resource metrics will be be relatively easy to implement and will provide the basis for transparency and strong decisions in food and energy policy.

- **Water management.** The U.S. should create a water czar whose responsibility includes monitoring and transparent reporting of water quality and availability. Aquifers should be mapped and their capacities monitored and reported. Over drafting should be budgeted and thoughtfully planned. The total energy cost of water transport should be calculated for every water district so policymakers can make informed decisions on subsidizing irrigation water. The Environmental Working Group led by Ken Cook has taken a strong first step by mapping the water costs associated with the Central Valley Authority.

  The National Corn Growers Association cites a USGS source that tells policymakers that corn production occurs on only 10% of irrigated land. A simple visual of the U.S. overlaid by corn ethanol distilleries suggests the USGS source may understate irrigated corn land by 300%.[225] Since every acre of irrigated corn consumes one million gallons of fresh water, the correct number of irrigated acres is critical for valid policy decisions.

- **Soil management.** The USDA might create a similar map showing various metrics on soil degradation including especially topsoil loss and salt invasion.

- **Air management.** The EPA should sponsor an independent study of air pollution caused by ethanol production and consumption. California might reinstate its suit against the EPA that shows a 26 – 30% crop loss for oranges, avocados and grapes caused by ethanol smog.

- **Energy management**. DOE and USDA might create metrics on food production and energy consumption so consumers could know the energy cost of their food. Food labels should reflect resource costs, especially water consumption, energy used to move the water and distribution costs.

- **Subsidy management.** Farm subsidies were established to support small farmers. Instead, subsidies currently create substantial social inequity because 80% of subsidy monies go to 10% of the huge agribusinesses. Subsidies enrich large farms so that they are able to buy out smaller farmers, magnifying social inequity.

  Agricultural subsidies promote monocropping which jeopardizes the food supply and the production of cheap, empty calories that threaten the health of our children. Four of the top 10 causes of death in America are chronic diseases linked to diet: heart disease, stroke, Type 2 diabetes and cancer.[226] The diets of the poor (and many affluent) lead to obesity and a litany of health disorders. The Centers for Disease Control estimate that one in three American children born in 2000 will develop Type 2 diabetes. Farm subsidies have decreased the cost of food at the appalling expense of doubling the cost of healthcare.

- **Health management**. More research needs to document the substantial health costs associated with mining, distribution and burning of coal in power plants. These health effects are critical because roughly 1/3 of the power grid energy is used to transport water and 80% of the water is used for irrigation in the U.S. Exposure to coal pollution creates environmental social inequity as people exposed to the coal plume not only face significant health risks but the loss of value of their land due to polluted air, soils and water. More studies are also needed on the effects of nitrogen pollution to surface and well water.

The USDA currently publishes strong metrics on crop production and the use of fertilizers, pesticides, herbicides and fungicides. However, considerably more research needs to occur on the health effects from those agricultural chemicals.

Production demonstration

Public demonstration of SAFE production will provide tremendous value because people could then visualize the simple process used to tap nature's oldest energy production system. Very few people have seen algal production systems in operation for three reasons:

1. Few outdoor algal production systems exist in the U.S. and those producers are typically distant from cities.

2. Producers carefully guard their intellectual property and trade secrets and do not give tours.

3. The federal government made a political decision to support ethanol research rather than algae in the 1990s, so there was no funding available for algal labs or R&D.

Two indoor and outdoor algae labs and production facilities at Arizona State University provide a typical example. Both labs were built with state monies and hosted tours that excited students, faculty and the community. Commercial enterprises made relatively small investments in the research and closed the labs and production facilities to the public – claiming intellectual property protection.

Greenindependence.org in collaboration with the Desert Biofuels Initiative, the city of Phoenix and other community groups is pursuing funding for open source demonstration and research facilities that will benefit all of mankind. The Phoenix demonstration facility is planned for City of Phoenix donated land that will be jointly used as a Fire Department training facility. Algal biomass will be produced using several different small-scale outdoor production models that provide visual interest and engaging education for visitors.

Several other cities are planning demonstrations sites that will provide research and education on SAFE production. Every city with a waste treatment plant offers great potential for co-locating green solar production systems to remediate municipal wastewater.

Public access to information

Currently, algal production knowledge rests in the brains of a select few people. While there are many phycologists around the world working in algae labs, there have been very few algal biomass producers in field settings. Restricting public knowledge further, much of this production knowledge is sequestered as intellectual property.

The algal industry today is fractured into a series of vertical markets with each firm acting as if they are a fiefdom where they must protect their intellectual property behind a wall of secrecy. Scientists from most firms are forbidden by non-disclosure agreements to collaborate or even talk honestly to other scientists about their projects. Their presentations at conferences are dumbed-down to protect their firm's intellectual property. Researchers are forbidden to share their production breakthroughs, costs or productivity metrics. Scientists are embarrassed about their limitations in social situations and even informal conversations. Information concentration leads to mistakes in algal production that are repeated multiple times, which has been fatal for several emerging firms. Firms cannot admit or share their mistakes for fear that the next round of funding will dissolve. Concentrated knowledge severely limits innovation because breakthroughs depend on only a few brains rather than many.

To moderate the public access problem, the greenindependence.org site supports the Green Solar Energy Alliance, (GreenSea), a collaboratory with open source architecture available to everyone. The knowledge base is searchable and organized under the relevant areas for algal production, harvest, component extraction and products. Published research, papers, ideas and experience contributed to the greenindependence.org wiki address the critical limitations and obstacles in order to assure green solar production. The sticky wiki includes intellectual property harvested from dated patents related to algal production.

The Sandra Day O'Connor College of Law at Arizona State University is creating an intellectual property pool for the algal industry where members may contribute IP to the pool. Users will pay a small fee for use of the IP and additional fees based on production volume. Intellectual property pools have served several industries including chemicals and pharmaceuticals.

New technologies, especially in the areas of food and energy, are certain to create unanticipated and unintended consequences. Active scientific, social and political analysis of SAFE production will be important to minimize unintended consequences. Probably the strongest fear for algaculture is that it would host some form of disease such as salmonella, e-coli or other pathogen.

The key solution for assuring positive impacts will be creating metrics, monitoring production and reporting in a manner that assures the public, scientists and policymakers have access to the data. Great Britain has developed a set of food safety metrics that may provide a model for monitoring and reporting the quality of green solar production. GreenSea will coordinate the vital quality control function in concert with other public partnerships.

GreenSea engages people from across the world in a coordinated initiative to end hunger by 2050. Freedom from hunger can be accomplished using a SOLAR strategy.

Green Solar Alliance Strategy to End Hunger

Search, discover and diffuse the knowledge and capability of the best ways to design, build and operate green solar production systems that consume non-fossil resources.

Organize collaborative research and practice on green solar production systems that optimize productivity and minimize cost.

Listen and learn the needs of hungry and needy people globally and transform their desires into practical sustainable and affordable food and energy production they can practice locally.

Assist people communities and governments globally to educate, plan, build, operate and maintain SAFE production systems.

> **Revolutionize** food and energy production by enabling people globally to grow SAFE production locally.

GreenSea democratizes access to green solar energy and plans to endow Green Masterminds with local control over production. Green Masterminds are similar to a Master Gardeners except that instead of knowing a lot about a wide range of plants, Green Masterminds will know enough about one plant, algae, to grow it productively. The process facilitates the creation of community cooperatives to produce, extract and process green biomass for local needs.

GreenSea operates as a collaboratory so the goals evolve with the group consensus process. The focus remains on sustainable, carbon neutral, environmentally sensitive solutions for SAFE production and follows these principles.

Table 10.1 Green Solar Alliance Principles

Principle	Meaning
Sustainable and ecologically positive.	• Sustainable and renewable for multiple generations. • Environmentally positive — improves rather than degrades air, water or soils. • Carbon neutral — emits no or minimal fossil CO_2 to the atmosphere. • Carbon capture — consumes, sequesters or recycles CO_2. • Net energy positive — produces more energy than it consumes. • Habitat positive — does not destruct or degrade valued conservation areas.
Consumes no or minimal	• Traditional agricultural crops have already taken most the best cropland for decades.

cropland, freshwater or fossil fuels.	• There is insufficient freshwater to irrigate needed land-based food crops for more than another two decades. • Peak fossil fuels means the cost of fuels will increase beyond the practical capability of food crop inputs.
Produced locally all over the world because climate change, poverty and hunger are global challenges.	• Produces food and fuel independent of altitude, latitude, longitude, geography or climate. • Produces food and energy in remote rural areas, small farms, villages, towns, cities and urban slums. • Produces independent of country, potentate, language, sociology, religion or politics.
Produces food and energy plus valuable coproducts, including:	• Good food and nutrients for people. • Fodder for animals, poultry and fish. • Water and air remediation — cleans polluted water and air. • Organic foods, nutrients, ingredients and health foods for people. • Organic fertilizers, biodegradable plastics and cosmetics. • Fine medicines, pharmaceuticals, nutraceuticals and vaccines.
Social equity	• Enables those with the least means access to affordable inputs for food and energy • Moderates a host of issues that plague the web of poverty and hunger

| **Appropriate technology** | • Fits with the local level of knowledge |
| | • Aligns with ethical considerations and local politics and social issues |

These SAFE production principles make sense in an environment plagued with fossil and human pollution, severe challenges to human health, peak oil and global climate change.

Social and political drivers

A variety of economic, social and health factors will accelerate green solar production. The rising cost of fossil fuels and foods, while bad for consumers, will prompt more R&D and heighten consumer interest in SAFE production. As scientists and journalists create a clear link between crop subsidies that create monocultures and the resulting health impacts on consumers, Congress will be motivated to shift farm policies to sustainable, ecologically positive and healthy crop subsidies.

The Sierra Club, Worldwatch, Greenpeace, Audubon Society, World Future Society and many other environmental and social organizations are promoting green initiatives that help consumers understand the value proposition for SAFE production. Scientific organizations such as the Union of Concerned Scientists, the Environmental Working Group and The National Energy Independence Plan, NEI, are providing the foundation for change in national energy policy. Strong voices including Al Gore, T. Boone Pickens, Bill Bailey, NEI, Ken Cook, ERG and many others help move public opinion towards sustainable and renewable energy solutions.

Some communities are orchestrating the systematic collection of used restaurant cooking oils and are home-brewing biodiesel.[227] Arizona's Desert Biofuels Initiative's "Gold to Green" project, led by Brad Biddle, Eric Jones and Sam West hopes to refine every drop of used restaurant cooking oil in Arizona to green diesel and remove 100 tons of fossil fuel pollutants from Arizona's air each year.[228]

Flipping point

For decades the cost of algal oil production has been a high multiple of fossil fuels. Much of the cost of fossil fuel production has been hidden from consumers by two factors: extraordinary government subsidies to Big Oil and completely ignoring the externalities associated with fossil fuels. The external costs of fossil fuels include carbon loading of the atmosphere which causes global climate change and carbon loading of the oceans which increases acidity and destroys coral reefs and shellfish. Possibly the highest external cost of fossil fuels are the health effects from mining, transporting and burning coal and petroleum.

When the total cost of fossil fuels is calculated, including the ecological, social and health, cheap fossil energy production becomes far more expensive. Consumers will call for SAFE production that has positive ecological, social and health impacts. However, until algal oil becomes directly cost competitive with gasoline, politicians and consumers are likely to choose the dirty option. When algal oil can be produced and delivered at prices similar to fossil gasoline, the energy industry will flip from fossil to clean renewable algal oil.

Interest in algae as a biofuel source began during World War II and German scientists were able to work out basic production models. However, the lack of biotechnology knowledge combined with the constraints of war doomed production. In the 1990s, several countries including the U.S. examined algae production for fuel but investment in algal R&D made no sense when fossil oil was $10 a barrel.

The recent interest in algae for liquid transportation fuels has been sparked by the dramatic drop in the cost of production as shown in Figure 9.2. Industry projections for 2009 indicate production prices under $10 are likely. The Defense Advanced Research Projects Agency, DARPA, recently let an RFP asking for algal oil production for jet fuel to be produced local to military bases for $3 a gallon, Figure 9.1. While that number is unlikely in the near future, scientists believe $6 a gallon local production jet fuel is feasible within five years.[229]

Demonstrated laboratory production capabilities combined with substantial price reductions in industrial production motivated global investments in algal production that exceeded $300 million in 2008. Sapphire Energy, the biggest winner to date, received a $50 million dollar investment in R&D from the Bill and Melinda Gates foundation and additional investment from the Rockefeller foundation.[230]

Figure 10.1 Price of Algal Oil per Gallon – Industrial Production

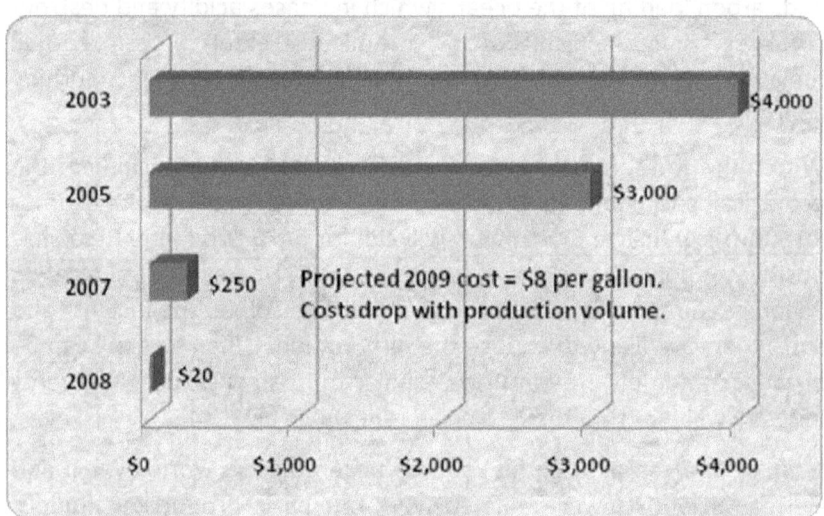

Public-private partnerships, investments between governments, universities, research labs and private companies, (DARPA, NREL and The Carbon Trust) have received substantial investments including:

- Exxon invested $600 million in Synthetic Genomics.
- Sapphire Energy received $50 million each from the Gates and Rockefeller Foundations.
- Chevron partnered with National Renewable Energy Laboratory (NREL).
- Aurora Biofuels received a $20 million second round of funding.
- Solazyme completed a $50 million round of funding.
- British Petroleum and the University of California at Berkeley and Arizona State University.
- The DOE invested $2.3 million in algae projects.

- Shell has invested with Cellana in a joint venture with the Hawaiian Natural Energy Laboratory and HR Biopetroleum.
- UK's Carbon Trust announced the world's largest algae challenge ($40 million).

Investments in private algae companies include the considerable monies flowing to Sapphire Energy. Investments in the first stage of commercialization of algae for fuel also received monies although commercial production of algae for fuel has yet to be proven viable. Several large producers are currently building plants and expect production in 2009.

Successful algal biofuel production will lead to exponential growth in the algal industry which will also fuel expanded R&D. These activities will broaden knowledge of algal production and provide new models for every element in algal production.

Is abundant agriculture possible?

Naysayers are quick to point out that others have proposed algae as a solution for world hunger but their initiatives have crashed and burned. The Internet is filled with wild speculation, both positive and negative, from people with little or no experience in algal production. To provide a credible answer, an Algal Industry Survey was developed to examine the critical industry issues from the perspective of industry leaders. Survey responses came from two sources:

1. The Algal Biomass Summit which was held in Seattle during October 2008 sponsored by the Algal Biomass Organization and yielded 46 respondents.
2. Algae World 2008 which met in Singapore in November and yielded 137 respondents.

Only a few questions will be reported here. The full survey report is available at the greenindependence.org site.

Algae Biomass Summit respondents were predominantly from North America were presented with two related questions: "If the U.S. focused on algaculture, how soon could the industry replace corn ethanol? About 9 billion gallons of ethanol will be produced in 2009." A similar question followed for oil imports.

Replace 9 B gallons of ethanol	Summit
10 years	63%
11 – 20 years	27%
never	10%
Displace 50% of oil imports	**Summit**
10 years	21%
11 – 20 years	57%
never	22%

Most respondents believe that a focused algaculture program could replace corn ethanol production in 10 years and 50% of oil imports within 20 years. About one in five do not believe algae can displace oil imports.

Algae World 08 respondents were presented with the question: "If your country focused on algaculture and made it a top priority, how soon could the industry replace 100% of imported oil?"

Replace 100% of imported oil	World
10 – 20 years	47%
21 – 40 years	34%
never	19%

International respondents were generally optimistic regarding the probability of replacing imported oil with home grown algal oil within 20 years. Several narrative comments indicated that replacing oil imports was their top priority.

The fact that U.S. and international producers believe they can produce significant amounts of algal oil for biofuels means similar production capacity could be available for food production. The following two questions requested narrative comments and were also informative.

What three key things need to happen to move the algal industry forward?

The dominate words used to enhance the industry were information, financing, funding sources, education and production. The industry needs government investment because the initial investments and risks are very high. The recommendations are grouped and ordered based on how often respondents recommended the action.

Information. The industry needs information sources to convey the algae story.

- Why isn't there a website that summarizes key algal sources, links and resources?
- Is there a profile of businesses and what they're doing in the algal industry?
- Can we create information sheets, pamphlets and books to inform people about algae?
- Can we create a PR campaign, a political action committee and lobbyists?

Financing. The industry needs considerably more funding from both public and private sources.

- Where are existing firms getting their funding?
- How can the algal industry persuade government to provide more R&D funding?
- How can investors evaluate algal investment opportunities?

Education and training. The industry needs to find ways to train people in algal production.

- Where can algal companies find trained technical and professional personnel?
- What kind of algal training is available and where?
- What technical skills and competencies are critical for building an algal industry?
- Is there a grade school and college curriculums for algae?

Production. The industry needs to describe in define algal production.

- What are the key production issues and how are they solved?
- What is the level of investment necessary for algal production?
- How can production losses be minimized from invasive species or grazers?
- What are practical methods for algal extraction and biomass processing?
- Can small community sized production systems be developed?

Collaboration. The industry needs to find ways to enhance networking and shared knowledge.

- How can information be shared and still maintain appropriate intellectual property protections?
- What kind of collaborative social network might work for algal professionals?
- Is there a summary of bloggers and journalists who follow algae?

What algal information would most benefit you?

Respondents indicated a desire for better information on algae, insights on financing, support for education and training, production information and stronger collaboration. Additional requests included the ideas below.

- **Industry summary.** The algal industry may follow the lead of other forms of renewable energy such as solar and wind to create a summary of the industry.
- **Demonstration units.** The algal industry should build and operate demonstration facilities so that people can see algal production.
- **Decision support.** Information on production, extraction and processing are too distributed and need to be more accessible.
- **Products and coproducts.** What is the total product array for algal biomass and what production strategies are used to maximize each product?
- **Real production numbers.** Actual rather than theoretical production numbers would be a huge breakthrough.

- **Ideal strains.** What are the ideal strains for various products and what are sources for these strains?
- **Market trends.** What are market trends in the algal industry?
- **Independent reviews.** Are there independent reviews of algal production methods?

Most industry participants believe algal production will focus on three biofuels; green diesel, gasoline and jet fuel, (JP-8.) There was no industry consensus on a best approach to algal biomass production including growing systems or production locations. Algal producers are experimenting with a diverse set of production models. Production models seem to vary based on the production objectives, type of feedstock and location. International producers tend to use open ponds while U.S. producers are planning to use closed or semi-closed CAPS. International producers are using naturally selected algae species while U.S. producers are planning to use a combination of species selection and genetically modified organisms that maximize the production of algal oil. These insights will provide guidance to leadership as they build the nascent algal industry.

Game changers

Breakthroughs to any of the SAFE production obstacles identified in the prior chapter will advance green solar farming, especially finding ways to lower costs and simplify production technology. Innovations that will accelerate adoption include the following.

- **Success models.** The major impediment to SAFE production comes from the lack of successful models. When modern farmers are able to co-locate reliable and affordable CAPS to support their land-based farms, diffusion will move quickly through global collaboration.

- **Home brew.** When home gardeners can create beer or wine, an industry will spring up faster than algae grow.

- **Lasagna gardens.** When gardeners can grow biomass effectively in the top of their solar garden, fish in the middle and shell fish on the bottom, many gardeners will become meat producers.

197

- **Mixed growth gardens.** Taking the lasagna garden another step, mixed use gardens may add aquaculture to produce additional vegetables, grains or fruit.

- **Compost plus solar gardens.** When a simple method like composting organic waste provides a rich organic nutrient supply for algae, gardeners would have a cheap source of nutrients.

- **Zoos and botanical gardens.** Demonstration algal production gardens in zoos to recover, recycle and reuse ZooPoo would educate millions of visitors to the importance of waste stream energy and nutrient recovery. ZooPoo would provide considerable feed for a wide variety of zoo animals and fish. Demonstration facilities would also be great for aquariums and botanical gardens.

- **Prisons.** The U.S. currently has twice as prisoners as farmers. Prisons consume extensive resources and create a large waste stream. CAPS could engage trustees in green collar jobs, remediate waste streams, produce algal biomass for tilapia and other fish as well as organic fertilizer for local fields.

- **Special needs farms.** Research on special needs kids and older people with dementia and Alzheimer's provides compelling evidence of vitality, health and cognitive improvement that occurs from animal and plant husbandry. Special needs farms could integrate green solar into their food production as well as their energy and sustainability educational systems.

- **CO_2 recovery.** If an effective means to recover CO_2 from vehicle exhaust pipes or other sources were developed, solar farmers would have a cheap source of carbon.

- **Mass production.** When CAPS are mass produced and cheap, people will supply their energy and ingenuity for production. The same applies to the pipes, valves, monitors, mixers and other necessary equipment. A cheap source of biodegradable plastic may evolve as a coproduct of the algal biomass for fuel industry.

- **Toxin control.** Finding cheap, low tech yet effective ways to control toxins and other microorganisms will be a huge benefit.

- **Clean water strategy.** When a country such as Zimbabwe recognizes that green solar can remove the the bacterium *Vibrio cholera* that causes cholera and parasites that cause a host of other illnesses from water supplies, they will promote the adoption of green solar, assuming they can find funding.

- **Micronutrient strategy.** When a country lacking in a micronutrient such as Vitamin A, B, C, D or iodine decides green solar could provide a cheap yet effective solution, government policies will spur diffusion. Algae are able to concentrate micronutrients even when local people are malnourished and nutrient deficient because algal biomass can bioaccumulate micronutrients that are in very low dilution in the local water.

- **Vaccine or medicine strategy.** When a country plagued by malaria or HIV/AIDS decides solar farms may provide a cheap yet effective solution, roll-out will occur quickly.

Most likely, several excellent breakthroughs have already occurred yet no one has thought to apply the new technology to algal production. Building the GreenSea collaboratory that systematically searches for novel solutions and recognizes and rewards green solar innovators will advance demonstration and diffusion. The collaboratory will also convey green solar's value chain.

Green solar value chain

When the details of algal production are resolved, hungry people may share in Algae's Green Promise for SAFE production. Algae are uniquely positioned to provide a value chain of products and solutions for critical human needs, Figure 9.2. The value chain includes sustainable and affordable foods for people, fish, fowl and animals. Algae can provide liquid transportation fuels that displace fossil fuels enable countries to become oil independent. Algae provides clean burning oil for cooking and heating fires that can end smoke death for millions of mothers and children who inhale black smoke particulates as they cook over wood, coal or dung.

Figure 10.2 Green Solar Products and Solutions

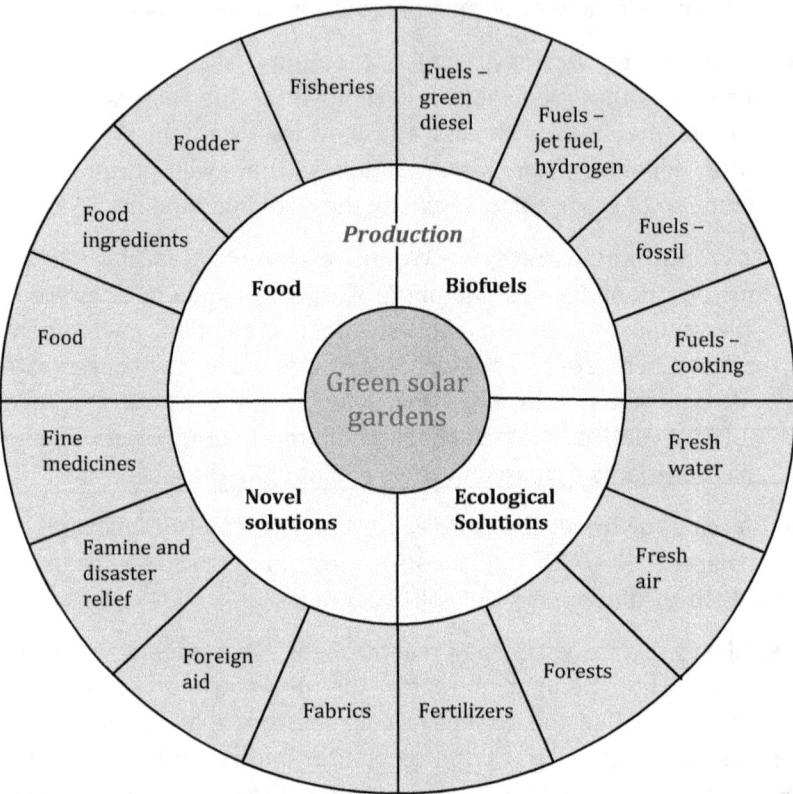

Algae's ability to clean waste and brine water offer a high-value solution for both the developing and developed societies. Israel currently recycles 87% of its municipal water through algal ponds to clean the water. Coal fire power plants and industrial manufacturers can flue their chimney smoke through algal ponds to sequester the CO_2 and heavy metals. Algal biomass production in villages can provide not only freshwater but fodder for grazing animals and reduce overgrazing which will save grasslands and forests from being denuded. Algal carbohydrates can be used to produce paper which can also save forests.

Many farmers in developing countries cannot afford fossil fertilizers. Green solar production can provide rich fertilizers without the use of fossil fuels that poor farmers can use directly on their land. Algal fertilizer also provides rich organic matter that builds humus and aids in water retention while protecting soil from erosion.

Algal biomass is not a full solution to hunger because the plant is low on calories, at about 2%. However, saving forests from grazing animals will allow communities to plant legumes and nut trees that can provide needed calories.

Knowledge and capability transfer for SAFE production will enable communities to end their dependence on food distribution. Teaching people to grow their own food and energy makes a lot more sense than a foreign-policy or disaster relief based on creating dependence by gifting food. As more green solar producers gain experience, many new innovative products and solutions will appear.

Algae's Green Promise, described in more detail in Table 10.2, will deliver an extraordinary green bounty while consuming no fossil resources and providing a positive ecological footprint. Green solar will not replace traditional fossil agriculture but will offer a new source of food and energy using agriculture based on abundance.

Table 10.2 Algae's Green Promise

Food

1. Food. Algae supply high-protein, low fat, nutritious, healthy and delicious human foods. Algae provide more vitamins, minerals and nutrients than land plants and are a natural health food. Algae are more than 50 times more productive than land plants but are not a full solution for malnutrition due to their low caloric value.

Note: Algae's food value will be suboptimal until solutions are found for making hard cell walls digestible. Only about 24 strains with digestible cell walls are available. All other green promises await only macro and micro-scale algaculture systems.

2. Food ingredients. Algal ingredients already enhance more than half the food, cosmetic and medical products in a grocery store. Algae components support dairy products, beer, jams, bakery products, soups, sauces, pie fillings, cakes, frostings, canned vegetables, colorings, ulcer remedies, digestive aids, eye drops, dental creams, skin creams and shampoos.

3. Fodder. Algae produce high-protein, low cost, nutritious animal feed with numerous vitamins, minerals and nutrients. Local production in villages could feed millions of animals and could save 20 million acres a year of forests and grasslands from desertification due to overgrazing.

4. Fisheries. Algae provide high-protein, low cost, nutritious fish feed with extensive vitamins and nutrients. Algae can be grown *in-situ,* in the water with the fin fish or shell fish. Fish tend to grow with more vitality on algae than land grains because many fish eat algae directly or some of the higher tropic organisms that eat algae in their natural habitat.

Fuel

5. Fuels – biodiesel. Algal oil from algal biomass produce renewable and sustainable, high energy biofuel such as clean diesel from an abundant and cheap stream of natural inputs.

6. Fuels – jet fuel, ethanol and hydrogen. Algal oil may be refined to a variety of high energy liquid transportation fuels including gasoline, ethanol or jet fuel. Algal products may be refined in existing refineries and made into products made from crude oil including biodegradable plastics.

7. Fossil fuels. Algal oil production on large green solar farms can displace fossil fuel imports which are 97% liquid transportations fuels. Algal oil uses recycled CO_2 and saves the costs and pollution associated with burning fossil fuels.

8. Fire – cooking. Black smoke from cooking fires and heating with wood, weeds and dung causes smoke death for 1.6 million and disability for 10 million mostly women and children every year. Clean burning, high energy algal-oil can end smoke death and the many smoke disabilities. Substituting algal oil for wood and agricultural materials will save a tremendous amount of labor from gathering firewood and allow forests to survive.

Ecological Solutions

9. Fresh water. Running animal and human wastewater through solar gardens feeds the plants nitrogen while the algae clean the water. Producing fuel, fodder or fertilizer using waste or brine water recovers water that can be used for land

10. Fresh air. Flueing effluent gasses from industrial, power or other smoke stacks through solar gardens removes CO_2, nitric oxides, sulfur and heavy metals such as mercury. Algae sequester greenhouse gasses; cleans the air and releases pure oxygen.

11. Fertilizer. Nitrogen fixing algae may provide high nitrogen, potassium and phosphorous fertilizers at very low cost in both production and energy inputs. The organic fertilizer could provide cheap local fertilizer and soil conditioner to subsistence farmers globally. The ash retains fertilizer value after being burned in cooking fires.

12. Forests. Algal biomass can end the need overgraze grasslands and forests by providing highly nutritious fodder for animals. Families will not need to denude forests and grasslands for cooking and heating fires because algal oil can serve those purposes. Villagers may replant their forests with nut trees or legumes for food to offset the low calories provided by algal foods.

Novel Solutions

13. Fabrics. Algal carbohydrates may be made into textiles, paper and building materials. Algal paper and building materials save forests and fabrics and provide warmth. Algal oils offer long chained hydrocarbons that can be made into anything refined from crude oil including biodegradable plastic.

14. Foreign Aid. American foreign aid provides subsidized U.S. food that undermines or destroys local food production because local farmers cannot compete with U.S. subsidized food. Gifting food fails to address the root cause of hunger or the web of hunger and poverty. An effective foreign aid strategy would be to transfer knowledge, capability and some start-up materials to grow SAFE production in the settings where people have need.

15. Famine and disaster relief. Algae, with its rich set of vitamins and minerals, may activate the immune system and stave off starvation while providing fuel, fodder, fabrics, fertilizers and fine medicines. Disaster relief with local SAFE production may prevent community starvation for millions. Local algal production solves the critical problem of food distribution and sustained food supply.

16. Fine medicines. High quality, low cost medicines and pharmaceuticals may be made from algal coproducts or grown in bioengineered algal strains to produce advanced compounds such as antibiotics, vitamins, nutraceuticals and vaccines. These compounds are grown today in land plants and animals so algae offer significantly faster, easier and lower cost production.

Designer algae grown locally in villages could save millions of lives by providing low cost vaccines or other medicines that need no packaging or distribution. Children or adults could simply eat the algae and benefit from the vaccine.

Fine medicines, especially personalized drugs tailored to an individual may offer more value than all other algal coproducts combined.

Nature's first food production system on Earth, algaculture, offers a wide range of potential benefits.

> *Things alter for the worse spontaneously, if they be not*
> *altered for the better designedly.* — **Francis Bacon**

Our collaborative task is simple: design, build and implement SAFE production systems.

Our shared path forward

Human societies currently exploit roughly half of the total primary biological production of the planet – through production of food, fisheries, forests, fabrics, fine medicines and other activities. Nearly all productive natural ecosystems on continents and oceans have been modified, manipulated or mined to support human consumption. Industrial agriculture has increased food production and enabled significant population expansion that now depends on fossil resources that are diminishing in supply and increasing in cost. Modern agriculture has accelerated the extraction of non-renewable fossil resources in a manner that is unsustainable. Global ecosystem exploitation will collapse with the end of available freshwater, fertile soils and economically recoverable fossil resource reserves required for food production.

Human societies will survive or perish in the near future based on our ability to manage the Earth's remaining biological and physical assets. Societies must curb population expansion, learn to recycle and reuse resources and practice sustainable consumption. The 2008 food riots in 40 countries where only a harbinger of the severe social unrest to come and served as a clear signal that action is needed urgently.

The key to human survival may be designing, developing and diffusing abundant agriculture built on a secure foundation of plentiful and inexpensive resources that will not crash.

Acknowledgements

Thanks first to my best friend and life partner, Ann Ewen, who made this project possible. Ann supported both the "ah's" and the "aha!'s"

This product would not have been possible without the extraordinary research of Lester Brown, President of the Earth Policy Institute and Jeffery Sachs of the Earth Institute. Professors Qiang Hu, Milton Sommerfeld and Bruce Rittman supported questions on molecular biology and algae production. Environmental scientists Al Darzins, Eric Jarvis and Mike Siebert at NREL were very helpful with renewable energy sources. Oilgae author Narasimhan Santhanam served with great sources. Thanks also to great advisors like Alina Kulikowski Tan who elevated *Green Algae Strategy* from a solo to an orchestra.

Science	Business – Econ.	Agribusiness
• Tom Moore	• Francine Hardaway	• Jon Ewen
• Jim Sears	• Mark Allen	• Gary Wood
• James Elser	• Alan Resnik	• Doug Young
• Andrew Ayers	• Susan Schultz	• Jim Robertson
• Herb Roskind	• Anthony Michaels	• Chris Kinsley

Also helpful were the published works of Paul Ehrlich, Sandra Postel, David and Marcia Pimentel, Nobel Laureate Al Gore, Harvey Blatt, Fred Pearce, Michael Pollan and Linda Graham. High-content websites were a great support such as Algaebase, U.N., W.H.O., the National Resources Defense Council, Sierra Club, Green Peace, Audubon Society, Union of Concerned Scientists, Center for Energy and Climate Solutions, Clean Water Network and Public Citizen. Also useful were U.S. government sources including: DOE, EPA, USDA, NOAA and NREL.

Mark Edwards

Mark has three goals he is pursuing through GreenIndependence.org:

1. End hunger
2. Free us from oil imports
3. Recycle our waste streams

These are urgent global challenges and we need to engage millions of Green Masterminds globally who have the capacity to grow algae for local needs.

Algae's green promise, capturing carbon while producing food and biofuel offers green independence from oil imports for America. Green Independence engineers hope for millions of global citizens who lack access to affordable food, feed, fuel and fine medicines.

Mark graduated from the U.S. Naval Academy in mechanical engineering, oceanography and meteorology where Jacques Cousteau motivated and mentored his interest in the oceans and global stewardship. He holds an MBA and PhD in marketing and consumer behavior and has taught food marketing, leadership, sustainability and entrepreneurship at Arizona State University for 33 years.

Mark served as CEO of TEAMS Intl. for 24 years, the software and assessment firm he founded based on his research on advanced assessment technologies, talent and leadership assessment. He served as lead consultant for more than 400 firms globally. He was retained by many U.S. departments and the military, including DOE, DOD, Special Forces and the National Labs.

Mark served as a Director for a Fortune 50 foods company and has done extensive R&D on new foods, sources and consumer behavior. He has consulted for Monsanto, Pioneer Seeds, DuPont, Nabisco, Quaker Oats, General Mills, Borden and many other agribusiness companies. He has worked with senior executives at 15 large U.S. oil and gas firms as well as British Petroleum and Saudi Aramco.

Great Green Reading

Food, energy and economics

Thomas L. Friedman, *Hot, Flat, and Crowded: Why We Need a Green Revolution – and How It Can Renew America*, Farrar and Giroux, 2008.

Lester R. Brown, *Plan B 3.0: Mobilizing to Save Civilization*, Third Ed., W. W. Norton; 2008.

Jeffrey D. Sachs, *Common Wealth: Economics for a Crowded Planet*, Penguin Press HC, 2008.

Jeffrey D. Sachs, *The End of Poverty: Economic Possibilities for Our Time*, Penguin Press, 2005.

Fred Krupp and Miriam Horn, *Earth: The Sequel: The Race to Reinvent Energy and Stop Global Warming*, W. W. Norton, 2008.

Brangien Davis and K. Wroth, *Wake Up and Smell the Planet: Non-Preachy Grist Guide to Greening Your Day*, Mountaineers Books, 2007.

Daniel Esty and Andrew Winston, *Green to Gold: How Smart Companies Use Environmental Strategy*, Yale University Press, 2006.

Water

Elizabeth Kolbert, *Field Notes from a Catastrophe: Man, Nature, and Climate Change*, Bloomsbury, 2006.

Sandra Postel, *Pillar of Sand: Can the Irrigation Miracle Last?* W. W. Norton & Company, 1999.

Peter H. Gleick, *The World's Water 2006-2007: The Biennial Report on Freshwater Resources*, Island Press, 2006.

Fred Pearce, *When the Rivers Run Dry: Water –The Defining Crisis of the Twenty-first Century*, Beacon Press, 2007.

Vandana Shiva, *Water Wars: Privatization, Pollution, and Profit*, South End Press, 2002.

Robert Jerome Glennon, *Water Follies: Groundwater Pumping and The Fate Of America's Fresh Waters*, Island Press, 2004.

---

[1] Please see Royal Tea for the 21 magic fossil resources.

[2] Pimentel, David and Marsha Pimentel. *Food, energy and society*, third edition, New York: CRC Press, 2008, 27.

[3] Ibid.

[4] WHO: 2004, World Health Report, http://www.who.int/whr/2004/annex/topic/en.annex_2_en.pdf.

[5] Black, Robert, Saul Morris, and Jennifer Bryce. "Where and Why Are 10 Million Children Dying Every Year?" *The Lancet* 361 (2003): 2226-34.

[6] Pimentel, David and Marsha Pimentel. *Food, energy and society*, third edition, New York: CRC Press, 2008, 27.

[7] Ibid.,12.

[8] Manning, Richard. *Against the grain: how agriculture has hijacked civilization*, New York, North Point Press, 38.

[9] Scherr, Sara and Sajal Sthapit. Farming and land use to cool the planet, in The State of the World 2009, the World Watch Institute, 2009, 37.

[10] Kolbert, Elizabeth. Field Notes from a Catastrophe: Man Nature and Climate Change. New York: Bloomsbury, 2006: 96-97.

[11] Diamond, Jared. Collapse: How Societies Choose to Fail or Succeed. New York: Viking Adult, 2004, 54.

[12] Pollan, Michael. Farmer in Chief, New York Times, October 9, 2008.

[13] Pimentel, D. and Giampietro. The Tightening Conflict: Population, Energy Use, and the Ecology of Agriculture", 1994, http://dieoff.org/page69.htm

[14] Ibid., 12.

[15] Personal interview, Jon Ewen, December, 2008.

[16] Personal interview, Marvin Morrison, July, 2003.

[17] Pimentel and Pimentel. *Food, energy and society*, 43.

[18] Pimentel, D. and Patzek, T. Natural resources research, 14;1, 65 – 76, 2005.

---

[19] Halweil, B. "Pesticide-Resistance Species Flourish." Vital Signs. Ed. L. Starke. New York: Norton, 1999: 124.

[20] "In Brief." Environment, Sept. 2001: 8.

[21] Brown, Lester. World Grain Harvest Falling Short by 54 M Tons: Water Shortages Contributing. Earth Policy Institute, 23 Nov. 2003, 3.

[22] Klein, Gary. The Water-Energy-Greenhouse Gas Connection, California Energy Commission www.bayareavision.org/initiatives/PDFs/GreeningInfill_120607_Klein.pdf

[23] Environmental Working Group, 2009.

[24] Sawyer, John. Natural gas prices affect nitrogen fertilizer costs, 2001.http://www.ipm.iastate.edu/ipm/icm/2001/1-29-2001/natgasfert.html

[25] USDA ERS, Fertilizer Use, 2008. http://www.ers.usda.gov/Data/FertilizerUse/

[26] Rosset, Peter. Lessons from the Green Revolution, FoodFirst, op ed., April 8, 2000

[27] USDA NASS, June 29, 2007 News Bulletin, http://www.nass.usda.gov

[28] Edwards, Mark. BioWar I: Why Battles over Food and Fuel Lead to World Hunger, Tempe: AZ, LuLu Press, 18. Note: assumes feedlot beef.

[29] Dias De Oliveira, Marcelo E., Burton E. Vaughan, and Edward J. Rykiel Jr. "Ethanol as Fuel: Energy, Carbon Dioxide Balances, and Ecological Footprint." *BioScience 55:7 (2005):* 593–602.

[30] *Linking Land Quality, Agricultural Productivity, and Food Security.* United States Department of Agriculture, Economic Research Service AER-823, 31. http://www.ers.usda.gov/publications/aer823/aer823e.pdf.

[31] http://www.organicconsumers.org/articles/article_1371.cfm

[32] Food Planet, Food summit blames trade barriers, biofuels, June 4, 2008, www.planetark.com/dailynewsstory.cfm/newsid/48626/story.htm

[33] WHO: 2004, World Health Report, http://www.who.int/whr/2004/annex/topic/en.annex_2_en.pdf. (10/1/04).

[34] Murphy, Annie. Bolivian farmers fueled up by soybeans, marketplace, American public radio, January 27, 2009.

[35] Edwards, Mark. BioWar , 2007, 209.

[36] Ibid.

[37] Calculated by Earth Policy Institute using economically recoverable reserve figures in U.S. Geological Survey, Mineral Commodity Summaries, January, 2007.

[38] Cohen, David. Earth's natural wealth: an audit, New Scientist, 23 May 2007, 34-41.

[39] Ibid.

[40] Ibid.

[41] Gleick, P. The World's Water 2001. Washington, DC: Island Press, 2000: 52.

[42] Ibid., 52.

[43] Diouf, Jacques. *Turning the Tide Against Water Scarcity*. Food and Agriculture Organization of the United Nations, Mar. 2007. http://www.fao.org/english/dg/oped/index.html.

[44] Pearce, 24.

[45] FAO, ResourceSTAT, electronic database, at faostat.fao.org/site/405/default.aspx, updated 30 June 2007.

[46] Matt Weiser, Feds Document Shrinking San Joaquin Valley Aquifer Sacramento Bee, July13, 2009.

[47] California Department of Water Resources, 2009. http://www.owue.water.ca.gov/agdev/

[48] Gorman, Steve. California farms lose main water source to drought, Reuters, February 20, 2009.

[49] Fred Pearce, "Asian Farmers Sucking the Continent Dry," New Scientist.com, 28 August 2004.

[50] Shiva, Vandana. The Suicide economy of corporate globalization, April 5, 2004, Znet. http://www.countercurrents.org/glo-shiva050404.htm

[51] 1,500 farmers commit mass suicide in India, The Independent, 15 April 2009. http://www.independent.co.uk/news/world/asia/1500-farmers-commit-mass-suicide-in-india-1669018.html?tsp=1

[52] Scherr, Sara. Farming and land use to cool the planet, in Sustainable Agricultural Development Strategies in Fragile Lands, 1994,34.

[53] USGS. http://www.usgs.gov/hazards/wildfires/

[54] Carle, David. Water in California. Berkeley: University of California Press, 2004: 150.

[55] Pearce, 30.

[56] U.N. Environment Program (UNEP), Global Outlook for Ice and Snow (Nairobi: 2007), p.103; J. Hansen et al., "Climate Change and Trace Gases," Philosophical Transactions of the Royal Society A, vol. 365 (15 July 2007), 1949–50.

[57] USGS, http://pubs.water.usgs.gov/ds220/

[58] Barlow, Paul Ground Water in Freshwater-Saltwater Environments of the Atlantic Coast, 2003. USGS..

[59] Durwood, M., R. M. Dixon, and O. F. Dent. Consumptive use of water by major crops in Texas. Texas Board of Water Engineers, Bulletin 6019, 1960.

[60] USDA NASS, 2003, Farm and Ranch Irrigation Survey, Vol. 3. Special Studies Part 1, AC-02-SS-1, 176.

[61] Ibid., 179. http://www.nass.usda.gov/Census_of_Agriculture/2002/FRIS/fris03.pdf

[62] Pimentel, D., Berger, D, Filiberto, et al. Water Resources, Agriculture and the Environment, Environmental Biology, Report No. 04-1. New York State College of Agriculture and Life Sciences, 2004.

[63] Hutson, Susan S., Nancy L. Barber, et al. "Estimated Use of Water in the United States in 2000." U.S. Geological Survey, 2005. http://pubs.usgs.gov/circ/2004/circ1268/.

[64] Ashworth, William. Ogallala Blue: Water and Life on the High Plains, New York: William Norton, 2007, 13.

[65] Ibid., 13.

[66] Ibid., 155.

[67] EPA, Assessed Waters, in 2000 National Water Quality Inventory, EPA-841-R-02-001, EPA, Office of Water, Washington, D.C., August, 2002.

[68] FAO, Meat and meat products, *Food Outlook* FAO, Rome, December, 2006.

[69] Steinfeld, H. and Shalonda, P. Old players new players, *Livestock report 2006*, Rome: FAO, 2006, 3.

[70] Steinfeld, H. Livestock's long shadow: environmental issues and options, Rome: FAO, 2006.

[71] Blatt, H. *America's Environmental Report Card: Are We Making the Grade?* Cambridge, Mass.: MIT Press, 2005, 75.

[72] Brown, L., R. Hindmarsh, and R. Mcgregor. Dynamic Agriculture Book Three. 2nd. ed. Sydney: McGraw-Hill Book Company, 2001: 84.

[73] Ask the Meatman, http://www.askthemeatman.com/yield_on_beef_carcass.htm

[74] Falk, Dean, "Fresh Water Needs for a Dairy." http://www.oneplan.org/Stock/DairyWater.shtml

[75] Pearce, When the Rivers Run Dry, 4.

[76] Nierenberg, Danielle. "Rethinking the Global Meat Industry." State of the World: 2006. Ed. The WorldWatch Institute. London: Norton, 2006: 24.

[77] USDA ERS Briefing, 2008. http://www.ers.usda.gov/Briefing/Corn/

[78] Edwards, Mark. Biowar I: Why Battles over Food and Fuel Lead to World Hunger, Tempe: LuLu Press, 2008, 117.

[79] Edwards, Mark. BioWar I: Why Battles over Food and Fuel Lead to World Hunger, Tempe: AZ, LuLu, 2008.

[80] Personal communication, Professor von Shalk, Israe, Nov 2008l.

[81] The Dalai Lama's Little Book of Inner Peace. London: Element Books, 2002.

[82] FAO: 2002, Restoring the Land, ttp://www.fao.org/inpho/vlibrary/u8480e/U8480E0d.htm.> (12/17/2002).

[83] Pimentel, David. Environment, Development and Sustainability. Dordrecht: Feb 2006. Vol. 8, Iss. 1, 119-137

---

[84] Pimentel, David. Food, energy and society, 29.

[85] Ehrlich, P. R. and A. H. Ehrlich. One with Nineveh: Politics, Consumption, and the Human Future. Washington: Island Press, 2004: 71.

[86] Sample, Ian. Global food crisis looms as climate change and population growth strip fertile land. The Guardian, 31 August 2007.

[87] Pimentel, David. Food, energy and society, 29.

[88] Linking Land Quality, Agricultural Productivity, and Food Security. United States Department of Agriculture, Economic Research Service AER-823, 31. http://www.ers.usda.gov/publications/aer823/aer823e.pdf.

[89] NRCS, 2009, Model Simulation of Soil Loss, Nutrient Loss and Soil Organic Carbon Associated with Crop Production, http://www.nrcs.usda.gov/technical/nri/ceap/croplandreport/croplandreportapp.html

[90] Pimentel, David and Pimentel, Marcia. Food, Energy and Society, 3rd Ed., New York, CRC Press, 2007, 201.

[91] Uri, N. D. "Agriculture and the Environment – the Problem of Soil Erosion," Journal of Sustainable Agriculture, 16;4, 71-91.

[92] Iowa Environmental Council, "Biofuels in Iowa, Policy Advisory Statement" (Des Moines, IA: 26 January 2006), 5.

[93] "Awful Weather We're Having." Editorial. The Economist, 2 Oct 2004.

[94] Depalma, Anthony. "Dire Climate Forecast Includes the 100-Year Flood, Once a Decade." New York Times, 11 July 2007, A7.

[95] DNR, "State of Iowa, Public Drinking Water Program 2006 Annual Compliance Report" (Des Moines, IA: June, 2007.

[96] USDA NASS, June 29, 2007 News Bulletin, http://www.nass.usda.gov

[97] FAO, 1999. http://www.fao.org/docrep/T1765E/t1765e0a.htm#negative

[98] WHO: 2004, World Health Report, http://www.who.int/whr/2004/annex/topic/en.annex_2_en.pdf. (10/1/04).

[99] "OECD Highlights Chinese Pollution." Financial Times, 17 July 2007. http://www.ft.com/cms/s/932c36ca-348c-11dc-8c78-0000779fd2ac.html.

[100] Oster, Shai. "In China, New Risks Emerge At Giant Three Gorges Dam," Wall Street Journal, August 29, 2007, A1.

[101] Ibid., A1.

[102] http://archives.cbc.ca/science_technology/measurement/topics/1572/

[103] Brown, J. and Foucher, S. http://graphoilogy.blogspot.com/

[104] Nedler, Chris. Oil exports may soon dry up, *The futurist*, March April, 2009, 6.

[105] Terry Tamminen, Lives Per Gallon: The True Cost of Our Oil Addiction (Washington, DC: Island Press, 2006), 60.

[106] Ibid., 62.

[107] Edwards, Mark. Biowar I. 2007.

[108] Ibid.

[109] United Nations Food and Agriculture Organization. *Sustainable Bioenergy: A Framework for Decision Makers*, 8 May 2007.

[109] DOE, http://www.energy.gov/energysources/fossilfuels.htm

[110] Ibid.

[111] DOE, http://www.energy.gov/energysources/fossilfuels.htm

[112] Union of Concern Scientists, http://www.ucsusa.org/clean_energy/coalvswind/c02d.html

[113] Ibid.

[114] Freese, Barbara and Steve Clemmer, Gambling with Coal: How Future Climate Laws Will Make New Coal Power Plants More Expensive, Union of Concerned Scientists, September, 2006.

[115] Chronic illness linked to coal mining, Science Daily, March 27, 2008.

[116] Pope, C. Arden. Fine-Particulate Air Pollution and Life Expectancy in the United States, New England Journal of Medicine, Jan 22, 2009, 360:376-386.

[117] bBNET, U.S. oil import bill to top $400 billion in 2008. http://findarticles.com/p/articles/mi_m0EIN/is_2008_March_7/ai_n24380 607

[118] Adapted from: Rittman, Bruce. Opportunities for Renewable Bioenergy: Using Microorganisms. Biotech, Bioengineering, 2008, 100: 203–212.

[119] Holden, Kevin. Chinese Air Pollution Deadliest in World, National Geographic News, July 9, 2007.

[120] Hubert, Peter, The Carbon Curtain, Forbes Magazine, Sept. 01, 2008

[121] Ibid.

[122] Cohen, David. Earth's natural wealth: an audit, New Scientist, 23 May 2007, 34-41.

[123] FAO, Global fertilizer supply to outstrip demand, Feb 26, 2008. www.fao.org/newsroom/en/news/2008/1000792/index.html

[124] Earth Policy Institute, Oil and Food: A Rising Security Challenge, Earth Policy Institute, 2005.

[125] Dias De Oliveira, Marcelo E., Burton E. Vaughan, and Edward J. Rykiel Jr. "Ethanol as Fuel: Energy, Carbon Dioxide Balances, and Ecological Footprint." BioScience 55:7 (2005): 593–602.

[126] Lawrence, Felicity . Not on the Label Penguin. 2004, 213

[127] ERS. www.ers.usda.gov/Publications/ages8965/ages8965fm.pdf

[128] National Center for Integrated Pest Management, India. http://www.ncipm.org.in/asps/DisplayFertilizers.asp

[129] FAO 2001.

[130] Amber Waves, Recent Volatility in Fertilizer Prices, March, 2009.

[131] Bradsher, Keith and Martin, Andrew. Shortage and price of fertilizer threatens to make more hungry, Intl. Herald Tribune, May 1, 2008.

[132] Pimentel, David. Food, energy and society, 29.

[133] Union of Concerned Scientists, Industrial Agriculture: Features and Policy, 2009.

[134] Cimitile, Matthew. Crops absorb livestock antibiotics, science shows. Environmental Health News, January 6, 2009.

[135] Huber, C., R. Baier, A. Gottlein, and W. Weis. Changes in Soil, Seepage Water and Needle Chemistry."Forest Ecology and Management 233 (2006): 11-20.

[136] Agricultural Resources and Environmental Indicators, 1996-97. Agricultural Handbook, No. 712. USDA, Economic Research Service, July 1997.

[137] USDA, ERS, http://www.ers.usda.gov/Data/FertilizerUse/

[138] Shwartz, Mark. Study Highlights Massive Imbalances In Global Fertilizer Use, Medical News Today, June 20. 2009.

[139] Smil, V . Cycles of Life. Scientific American Library, New York, 2000.

[140] Rich, Deborah. The case against synthetic fertilizers, San Francisco Chronicle, Saturday, January 14, 2006.

[140] DNR, "State of Iowa, Public Drinking Water Program 2006 Annual Compliance Report" (Des Moines, IA: June, 2007.

[140] Beeman, Perry. State will spend millions to improve water quality, PBEEMAN@DMREG.COM, MAY 10, 2009.

[140] USDA, ERS, http://www.ers.usda.gov/data/FertilizerTrade/mainpage.asp?ERSTab=2

[140] Sawyer, John. Natural gas prices affect nitrogen fertilizer costs, 2001.http://www.ipm.iastate.edu/ipm/icm/2001/1-29-2001/natgasfert.html

[140] MarineLinc.com. Ocean Freight Shipping Rates on the Rise, Monday, January 07, 2009.

[140] Vaccari, David. Phosphorus Famine: The Threat to Our Food Supply, Scientific American, June, 2009.

[140] Ibid.

[141] Matsui, 1997 Matsui, S., 1997. Nightsoil collection and treatment in Japan. In: Drangert, J.-O., Bew, J., Winblad, U. (Eds.). Ecological Alternatives in Sanitation. Publications on Water Resources: No 9. Sida, Stockholm.

[142] Mårald, 1998 Mårald, E., 1998. I mötet mellan jordbruk och kemi: agrikulturkemins framväxt på Lantbruksakademiens experimentalfält 1850–1907. Institutionen för idéhistoria, Univ Umeå.

[143] Brink, 1977. Brink, World resources of phosphorus, *Ciba Foundation Symposium* **13–15** (1977), pp. 23–48.

[144] Hugo, 1862 Hugo, V., 1862. Les Miserables, ch.323, A. Lacroix, Verboeckhoven & Ce.

[145] Florida Institute for Phosphate Research, http://www1.fipr.state.fl.us/phosphateprimer

[146] Lewis, Leo. Scientists warn of lack of vital phosphorus as biofuels raise demand, The Times, June 23, 2008.

[147] Smil, 2007 Smil, V., 2007. Policy for Improved Efficiency in the Food Chain, SIWI Seminar: Water for Food, Bio-fuels or Ecosystems? World Water Week 2007, August 12th–18th 2007, Stockholm.

[148] Smil, 2000 V. Smil, Feeding the World: A Challenge for the 21st Century, The MIT Press, Cambridge (2000).

[149] EFMA, 2000 European Fertilizer Manufacturers Association, Phosphorus: Essential Element for Food Production, European Fertilizer Manufacturers Association (EFMA), Brussels (2000) and FAO, 2004. The Use of Phosphate Rocks for Sustainable Agriculture Technical. In: F. Zapata (Ed.), Joint FAO/IAEA Division of Nuclear Techniques In Food and Agriculture, Vienna, Austria, R.N. Roy Land and Water Development Division, FAO, Rome, Italy.

[150] EcoSanRes, 2003 EcoSanRes, Closing the Loop on Phosphorus, Stockholm Environment Institute (SEI) funded by SIDA Stockholm (2003).

[151] IFA, 2006 IFA, 2006. Production and International Trade Statistics, International Fertilizer Industry Association Paris, available:http://www.fertilizer.org/ifa/statistics/pit_public/pit_public_statistics.asp (accessed 20 August 2007).

[152] Jasinski, 2008 Jasinski, S.M., 2008. Phosphate Rock, Mineral Commodity Summaries, U.S. Geological Survey <minerals.usgs.gov/minerals/pubs/commodity/phosphate_rock/>.

[153] EFMA, 2000 European Fertilizer Manufacturers Association, Phosphorus: Essential Element for Food Production, European Fertilizer Manufacturers Association (EFMA), Brussels (2000) and FAO, 2004. The Use of Phosphate Rocks for Sustainable Agriculture Technical. In: F. Zapata (Ed.), Joint FAO/IAEA Division of Nuclear Techniques in Food and Agriculture, Vienna, Austria, R.N. Roy Land and Water Development Division, FAO, Rome, Italy.

[154] Corell, H, Letter dated 29 January 2002 from the Under-Secretary-General for Legal Affairs the Legal Counsel addressed to the President of the Security Council, *United National Security Council, Under-Secretary-General for Legal Affairs The Legal Counsel, 2002.*

[155] Kvarnström et al., Urine Diversion: One Step Towards Sustainable Sanitation, EcoSanRes program,, Stockholm Environment Institute Stockholm, 2006.

[156] http://www.improve-your-garden-soil.com/phosphorus-and-soil.html

[157] www.potashcorp.com/learn_about_fertilizer/about/processes/potassium/page_1.zsp

[158] MacNamara, William. China eyes developed mine assets, Financial Times. London (UK):Jan 6, 2009, 5.

[159] Environmental Risks of Pesticides Versus Genetic Engineering for Agricultural Pest Control
Maurizio G. Paoletti, David Pimentel. Journal of Agricultural and Environmental Ethics. Guelph:2000. Vol. 12, Iss. 3, 279.

[160] Pimentel, David. Environmental and Economic Costs of the Application of Pesticides Primarily in the United States, Environment, Development and Sustainability. Dordrecht: 2005. Vol. 7, Iss. 2, 229-252

[161] Johnson, J. "Blowing Green." Chemical & Eng. News, 24 Feb. 2004: 7.

[162] Blatt, 81.

[163] Dooley, E. "Protected Harvest." Environmental Health Perspectives, May 2002: A237.

[164] Walsh, Edward. "EPA Stops Short of Banning Herbicide." Washington Post, 2 Jan. 2003: A14.

[165] Hayes, T.B., A. Collins, M. Lee, M. Mendoza, N. Noriega, A.A. Stuart, and A. Vonk. Hermaphroditic, Demasculinized Frogs After Exposure to the Herbicide, Atrazine, at Low Ecologically Relevant Doses. Proceedings of the National Academy of Sciences (99:5476-5480), 2002.

[166] National Academy of Science. Population Summit of the World's Scientific Academies. Washington, DC: National Academy of Sciences Press, 1994.

[167] Mathis Wackernagel et al., Tracking the Ecological Overshoot of the Human Economy, Proceedings of the National Academy of Sciences, vol. 99, no. 14, 9 ,July 2002, 9,266–71.

[168] National Academy of Science, 1994.

[169] The Principal Agglomerations of the World, http://www.citypopulation.de/world/Agglomerations.html

[170] Lewis, Mark. 21st Century Cities, Megacities Of The Future, Forbes Magazine, 06.11.07.

[171] Pimentel and Pimentel, 2007, 43.

[172] James Hansen et al., "Climate Change and Trace Gases," Philosophical Transactions of the Royal Society A, vol. 365 (2007), pp. 1925–54.

[173] Zoellick, Robert. High-Level Conference on World Food Security, Rome, World Bank, June 4, 2008. No: 2008/349/EXC.

[174] http://www.triplepundit.com/pages/un-rome-conference-mobilisatio-003225.php

[175] Today in Biofuels, Biofuels Digest, http://www.biofuelsdigest.com/blog2/2008/03/07/today-in-biofuels

[176] http://www.larouchepac.com/node/10736

[177] http://deltafarmpress.com/news/food-procudtion-0604/

[178] Annual Energy Report" (PDF). US Department of Energy, 2006-07.

[179] World Bank, World Development Report 2008: Agriculture for Development, October, 2007. http://publications.worldbank.org/ecommerce/catalog/

[180] Nicholas Stern, The Stern Review on the Economics of Climate Change, London: HM Treasury, 2006.

[181] Mean temps since 1970. ERS.

[182] Moskin, Jullia. Outbreak of Fungus Threatens Tomato Crop, *New York Times*, July 17, 2009.

[183] Mohan K. Wali et al., "Assessing Terrestrial Ecosystem Sustainability," Nature & Resources, October–December 1999, pp. 21–33.

[184] Lobel, David and Gregory Asner. "Climate and Management Contribution to Recent Trends in US Agricultural Yields." Science 299 (14 Feb. 2003): 1032.

[185] J. Larsen, Earth Policy Institute, published online 28 July 2006 www.earth-policy.org/Updates/2006/Update56.htm

[186] W. Easterling et al., in Climate Change 2007: Impacts, Adaptation, and Vulnerability, M. Parry et al., Eds. (Cambridge Univ. Press, NY, 2007), 976.

[187] Roosevelt, Margot. Climate change could put the heat on California crops, LA Times, July 22, 2009.

[188] Tankersley, Jim. California farms, vineyards in peril from warming, U.S. energy secretary warns, Los Angeles Times, February 4, 2009.

[189] Scherr, Sara Farming and land use to cool the planet, 31.

[190] World Bank, World Development Report 2008:

Agriculture for Development, World Bank, Washington, DC, 2007.

[191] Battisti, David and Rosamond Naylor. Historical Warnings of Future Food Insecurity with Seasonal Heat, Science, 8 September 2008,10.1126

[192] Pennisi,Elizabeth. Western U.S. Forests Suffer Death by Degrees, Science 23 January 2009: 447.

[193] Blatt, 75.

---

[194] Nierenberg, Danielle, Ed. Rethinking the Global Meat Industry. State of the World: 2006. The WorldWatch Institute. London: Norton, 2006: 24.

[195] http://www.epa.gov/oecaagct/ag101/demographics.html

[196] ERG, http://farm.ewg.org/farm/summary.php

[197] http://www.organicconsumers.org/articles/article_1371.cfm

[198] Food Planet, Food summit blames trade barriers, biofuels, June 4, 2008, www.planetark.com/dailynewsstory.cfm/newsid/48626/story.htm

[199] Edwards, Mark. Green Algae Strategy: Engineer Sustainable Food and Biofuels, Tempe: CreateSpace, 2008, 40.

[200] Badger, Emily. We gotta save them. Miller-McCune.com, July 15, 2009.

[201] MacKenzie, Debora, Caterpillar leg strikes West Africa, *New scientist*, 31 January 2009, 12.

[202] Yang, X. B. Soybean Brown Spot and Bacteria Blight. http://www.ipm.iastate.edu/ipm/icm/2003/7-28-2003/spotblight.html

[203] Why the Sudden Oak Death and Soybean Blight? http://www.jgi.doe.gov/sequencing/why/suddenoak.html

[204] Linda Graham, *Algae*, 32

[205] Gerwick,W.H. et. al. Screening cultured marine algae for anticancer-type activity. Journal of applied phycology, 1994, 6: 143-149.

[206] UPI.com. Oil drilling may help biomedical research, July 3, 2008. http://www.upi.com/Science_News/2008/07/03/

[207] Wenjuan Jobgen, et. al. Dietary L-Arginine Supplementation Reduces White Fat Gain and Enhances Skeletal Muscle, J. Nutr. 2009 139: 230-237.

[208] Martone et al. Discovery of Lignin in Seaweed Reveals Convergent Evolution of Cell-Wall Architecture. Current Biology, 2009; 19 (2): 169 DOI: 10.1016/j.cub.2008.12.031

[209] Edwards, Mark Biowar I. 2007, 60-71.

[210] Edwards, Mark. Unpublished survey research.

[211] This figure averages the estimates from seven independent algae scientists who are producing algae.

[212] Sachs, Jeffery. Common Wealth, 44.

[213] Rogers, E. Diffusion of Innovations, 5th Ed., New York: Free Press, 2003.

[214] FAO, seaweed as human food,
http://www.fao.org/docrep/006/y4765e/y4765e0b.htm

[215] Andersen, Robert. Ed., *Algal culturing techniques,* Phycological Society of America, Elsevier Academic Press, 2005.

[216] *Sahoo, Dinabandhu and Charles Yarish,* Mariculture of Seaweeds in Andersen, Robert A., Ed. *Algal culturing techniques*, Phycological Society of America, Elsevier Academic Press, 2005, 219 – 232.

[217] Oswald, W.J. and C.G. Golueke, Biological transformation of solar energy, *Advances in. Applied Microbiology, 2, 1960,* 223–262.

[218] *Borowitzka, Michael A.* Culturing Microalgae in Outdoor Ponds, in Andersen, Robert A., Ed. *Algal culturing techniques*, Phycological Society of America, Elsevier Academic Press, 2005, 205 – 219.

[219] Behrens, Paul W. Photobioreactors and Fermentors: The Light and Dark Sides of Growing Algae, in Andersen, Robert A., Ed. *Algal culturing techniques*, Phycological Society of America, Elsevier Academic Press, 2005, 189 -205.

[220] Linda Graham, *Algae*, 32.

[221] Edwards, Mark and Schultz, Clifford. Reframing agribusiness: moving from farm to market centric, Agribusiness: an international journal, June, 6;2, 2005, 27-43.

[222] Upton, Simon. *Avoiding the Wrong Solutions to the Wrong Problems.* Director, The Global Subsidies Initiative. Conference on Energy Security, Rushlikon, Zurich, 9 Mar 2007.

[223] "How ADM Makes a Killing on Ethanol." *New York Times*, 25 June 2006.

[224] Jefferson, Thomas. "Letter to William Johnson, 1823." *The Writings of Thomas Jefferson*. Ed. Andrew Libscomb. Washington DC: Thomas Jefferson Memorial Association, 1901: 450-51.

[225] Edwards, Mark. Green Algae Strategy, 2008, 84.

[226] Food Stamp Program Average Monthly Participation. USDA Food and Nutrition Service. February 2008.

[227]Huth, Hans. Biodiesel 101, 2008.
http://www.inkacola.com/greenbeat/soybenz/b101man/

[228] Eric Johnson and Brad Biddle founded the Desert Biofuels Initiative.
http://desertbiofuels.blogspot.com/

[229] Edwards, Mark. Algal Biomass Summit Industry Survey, 2008.

[230] 2008 the year of algae investment. Aquanticbiofuel.com.
http://aquaticbiofuel.com/2008/12/05/2008-the-year-of-algae-investments/

www.ingramcontent.com/pod-product-compliance
Lightning Source LLC
Chambersburg PA
CBHW071411170526
45165CB00001B/236